SENCKENBERG

SENCKENBERG
MONOGRAPHS

2 · 2025

Thomas S. Lechner & Madelaine Böhme

The early Late Miocene hominid locality Hammerschmiede (Bavaria, Southern Germany) – excavation, stratigraphy, and taphonomic insights

Senckenberg Gesellschaft für Naturforschung
Senckenberg Forschungsinstitut und Naturmuseum Frankfurt/M.

ISBN 978-3-510-61430-1 | ISSN 2943-7539

Contents

SENCKENBERG
MONOGRAPHS

2: 1–97
2025

The early Late Miocene hominid locality Hammerschmiede (Bavaria, Southern Germany) – excavation, stratigraphy, and taphonomic insights

Thomas S. Lechner[1*] & Madelaine Böhme[1, 2]

[1]Eberhard Karls University of Tübingen, Department of Geoscience, Sigwartstr. 10, 72074 Tübingen, Germany
[2]Senckenberg Centre for Human Evolution and Palaeoenvironment (HEP)
*thomas.lechner@uni-tuebingen.de

Manuscript submitted: 28. February 2025, revised manuscript accepted 22. July 2025

Abstract

The early Late Miocene fossil site of Hammerschmiede has been known for over half a century, but prior research primarily focused on taxonomy. Excavations by the University of Tübingen since 2011, with enhanced documentation and excavation techniques from 2017 onwards, have significantly increased both the quantity and quality of fossil finds, enabling detailed spatial and taphonomic analyses. The sedimentary sequence of the Upper Series lithostratigraphic unit at Hammerschmiede dips gently northward (<1°), linking it structurally to the Inclined Foreland Molasse. The site features seven fossil-bearing strata dated to between 11.62 and 11.56 Ma at the base of the Tortonian, respectively Pannonian stages. Current excavations are focusing on two fossil-rich fluvial deposits (HAM4, HAM5), representing distinct environments: a dynamic meandering river (HAM4) and a smaller, more stable rivulet (HAM5). Differences in vertebrate assemblages, microfossil distributions, and charcoal abundance suggest distinct ecological and environmental conditions between the two levels, reflecting variations in vegetation, fire regimes, and habitat structure. The HAM4 river was probably surrounded by a fire-prone, relatively sparse tree vegetation, providing some open space in the riparian landscape. In contrast, the HAM5 rivulet, which harbours both syntopic great apes *Danuvius guggenmosi* and *Buronius manfredschmidi*, may have flown through a more densely vegetated habitat. Both watercourses are characterised by organisms indicating clean and well-oxygenated waters (e. g., unionid mussels, salmonid fishes, giant salamanders), but geochemical data reveal redox-driven mineral alterations of fossil bones and plant remains. This was probably caused by a strong redox boundary at the sediment-water interface, induced by muddy bedload sediments, which are generally very fine-grained and range from silty clay (HAM5) to silty fine sand (HAM4). Lithic components coarser than sand are practically absent in the Hammerschmiede sequence. Observations of the exposed channel structures, supplemented by multiple orientation measurements on elongated objects and by taphonomic observations on possible individual skeletal strewnfields, uncover a meandering flow direction of the HAM4 river from southwest towards northeast and for the HAM5 rivulet from south to north. Biostratinomic studies further reveal complex bone accumulation patterns influenced by bone properties, flow dynamics, sediment composition, and biological activity. Skeletal strewnfields, including remains of mammals, birds, and turtles, suggest carcasses were introduced episodically, with some partial decomposition before fluvial transport. The HAM4 river experienced significant discharge fluctuations, affecting bone sorting and preservation. In contrast, the HAM5 rivulet exhibited lower turbulence and better-preserved skeletal associations.

ISBN 978-3-510-61430-1 | ISSN 2943-7539

The exceptionally high biodiversity of Hammerschmiede, including currently 151 vertebrate species among which 86 are mammalian taxa, suggests a resource- and nutrient-rich ecosystem. This site may therefore provide a rare opportunity to study Late Miocene biodiversity at high temporal resolution. The well-documented stratigraphy allows α-diversity reconstructions at a resolution down to centuries, bridging the temporal gap between fossil and extant ecosystems. Additionally, the site preserves a geodiversity archive of climatic and ecological signals across multiple timescales. These findings underscore the importance of high-resolution excavation and documentation techniques in palaeontological research.

Keywords

Miocene, taphonomy, Hammerschmiede, meandering river, excavation techniques

1 Introduction

During the Late Mesozoic (mid-Cretaceous), the convergent movement of the African-Adriatic (Apulian) and European plates initiated extensive tectonic processes that led to the formation of the Alpine mountain ranges – an event known as the Alpine (Alpide) orogeny (Schwerd et al. 1996). Throughout the Cenozoic, the progressive and intensifying continental collision resulted in the accumulation of predominantly continental deposits within an orogenic wedge, which gradually advanced northward into the European Plate. As the growing mountain range underwent crustal thickening, the added mass caused it to "sink" into the earth's lithosphere and underlying mantle due to density compensation. At the same time, the peripheral lithosphere to the north experienced flexural bending. The depth and width of the resulting foreland basin were governed by the flexural rigidity of the lithosphere (Schwerd et al. 1996). In the case of the North Alpine Foreland Basin – also referred to as the Molasse Basin (from the German *Molassebecken*, itself derived from the French term *molasse*, meaning "very soft") – this basin extends more than 1000 km east to west, spanning France, Switzerland, Germany, and Austria along the entire Alpine front (Schwerd et al. 1996). The basin reaches a width of up to 130 km, with its primary phase of subsidence occurring between the Late Eocene and Late Miocene (Lemcke 1988, Schwerd et al. 1996, Kuhlemann et al. 2002, Doppler et al. 2005, Kuhlemann 2007).

The basin structure functions as a sediment trap, continuously accumulating material throughout its evolution. These sediments primarily consist of erosional products and debris from the rising Alps to the south but also include material from the Bohemian Massif to the north. This ongoing sediment accumulation further contributed to the basin's subsidence.

Due to significant variations in basin depth, sediment sources, and water supply, the deposited sequences show substantial differences across the entire foreland basin. The basin is deepest near the southern orogenic front and gradually rises at a low angle toward the north, resulting in sediment thicknesses of up to 5000 metres in the southernmost areas (Schwerd et al. 1996). A borehole near Opfenbach (Bavaria, Allgäu region), 65 km southwest of Hammerschmiede, documented 3650 metres of Molasse sediments (Lemcke 1988; Fig. 1.0.1).

The sedimentary record of the North Alpine Foreland Basin can be divided into four major sedimentary phases, which are characteristic of southern Germany and the Allgäu region in particular (Steininger et al. 1989, Schwerd et al. 1996, Scholz 2016). Depending on the degree of basin subsidence and global sea level fluctuations, marine transgressions of the Paratethys Ocean periodically flooded the foreland basin (Schwerd et al. 1996, Scholz 2016).

The oldest sediments in the Molasse Basin date back to the latest Eocene and belong to the Lower Marine Molasse (*Untere Meeresmolasse*, UMM; latest Eocene – earliest Oligocene) (Schwerd et al. 1996, Doppler et al. 2005, Scholz 2016, van der Boon et al. 2018). These are followed by terrestrial fluvial sand, silt, clay, and marl deposits, forming the Lower Freshwater Molasse (*Untere Süßwassermolasse*, USM; Early Oligocene – Early Miocene). Another marine transgression then led to the deposition of the Upper Marine Molasse (*Obere Meeresmolasse*, OMM; Early Miocene). The final major phase consists of terrestrial, predominantly fluvial deposits of the Upper Freshwater Molasse (*Obere Süßwassermolasse*, OSM; late Early Miocene – Late Miocene).

Ongoing Alpine tectonic activity from the south (Austroalpine nappes; *Alpine Decken*, Fig. 1.0.1) resulted in the overthrusting of the lowermost Molasse sediments (*Überfahrene Molasse*). In the southernmost region, parts of these deposits were folded and steeply inclined, forming the Folded Molasse (*Faltenmolasse*, Fig. 1.0.1) and Inclined Molasse (*Aufgerichtete Molasse*), tilting northward (Lemcke 1988, Schwerd et al. 1996). This subdivides the asymmetric basin into a southward dipping northern wing of Foreland Molasse and a northward dipping southern wing of Inclined Molasse (Fig. 1.0.1).

During the Late Miocene, foreland basin subsidence stopped, and the area slowly began to uplift, probably due to isostatic rebound caused by increasing sediment erosion of the orogen (Schwerd et al. 1996). Finally, the uplift led to increased denudation and erosion of the superimposed Molasse sediments during the Pliocene and Pleistocene (in particular by glaciers).

According to Doppler (1989) and Doppler et al. (2005) the Upper Freshwater Molasse of western Bavaria is divided into four units. It starts with the oldest *Limnische Untere Serie* (Limnic Lower Series), followed by the *Fluviatile Untere Serie* (Fluvial Lower Series) and *Geröllsandserie* (Gravel-Sand Series), and ends with the youngest section of the *Obere Serie* (Upper Series), which includes the Hammerschmiede locality. This last phase is characterised by fine-grained fluvial deposits, which sometimes contain a considerable amount of carbonate. In addition to this, thin lignite seams also occur there. In regional geological terms, these argillaceous-marly to fine sandy deposits of the uppermost Upper Freshwater Molasse are also known as *Flinz* because of the reflections caused by the abundance of micas (Klein 1937, 1938, Kordiuk 1938, Klein 1939).

Due to the strong erosion since the Pliocene, the youngest sections of the Upper Series are not preserved everywhere. A hill ridge at the western margin of the Wertach valley, in the area of the Hammerschmiede clay pit near Pforzen (Allgäu region) possibly comprises the youngest deposits of Upper Freshwater Molasse preserved in Bavaria (Kirscher et al. 2016).

1.1 Research history of the Hammerschmiede locality

The Hammerschmiede fossil site is an active clay mine near the small village of Pforzen (Bavaria, Southern Germany). The clay pit comprises fluvio-alluvial floodplain deposits (clay, silt, fine sand, lignite) of the Northern Alpine Foreland Basin (Fig. 1.1.1).

The area around the municipality of Pforzen and Irsee has long been known for the occurrence of lignite, which was even mined underground during the 19th and 20th century, and reported since at least 1836 (Wiedenmann 2011). Carl Wilhelm von Gümbel (1861, p. 791) was the first to list fossil finds (plant and mollusc remains) from the vicinity of 'Irrsee' (today Irsee), unfortunately without a detailed description of the site. While coal mining repeatedly declined due to a lack of economic viability, clay began to be mined in this area in 1947 (Wiedenmann 2011). In the description of a stratigraphic column of a 31-metre-deep shaft at Hammerschmiede from February 1948, the first fossils were localised and mentioned alongside the coal in the form of invertebrate fossils – gastropods – as components found in some clay layers within the profile (observation by Dr. Heim 17.02.1948 to head miner Karl Zill, in Wiedenmann 2011). This historical profile reached an altitude of 657 metres above sea level and was thus situated below the current Hammerschmiede outcrop.

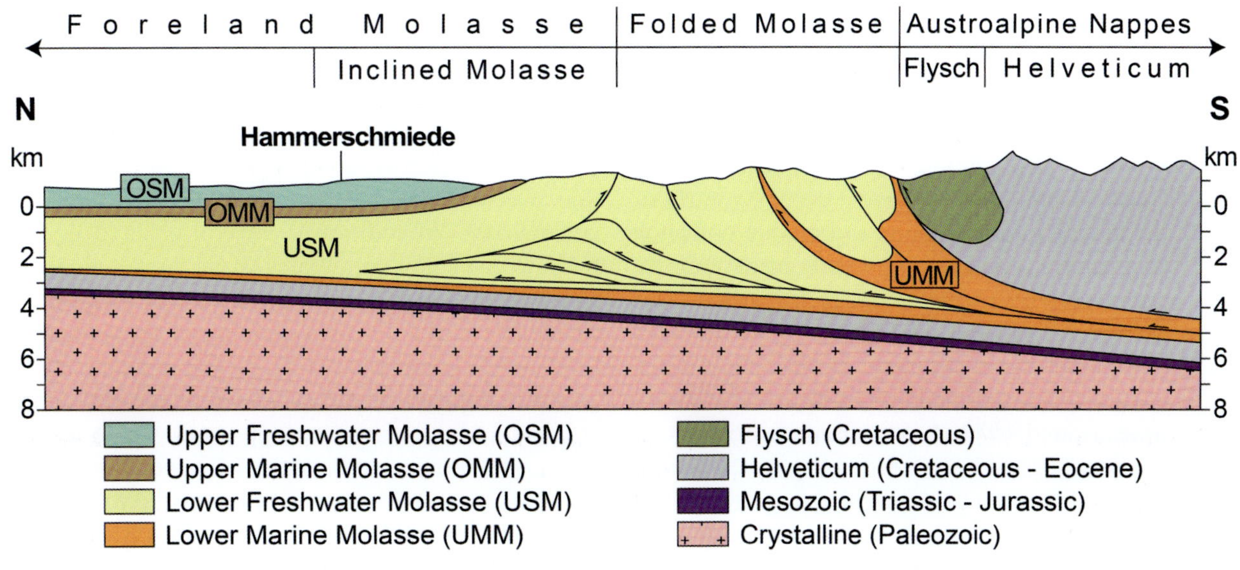

Figure 1.0.1. Schematic geological section through the subsurface of the Allgäu (southern Germany) and neighbouring areas (N–S). In the north, the undisturbed sediments of the Molasse basin, followed by the inclined southern margin, the folded Molasse, and the adjacent Alpine nappes of the Alpine orogen in the south. Modified after Scholz (2016).

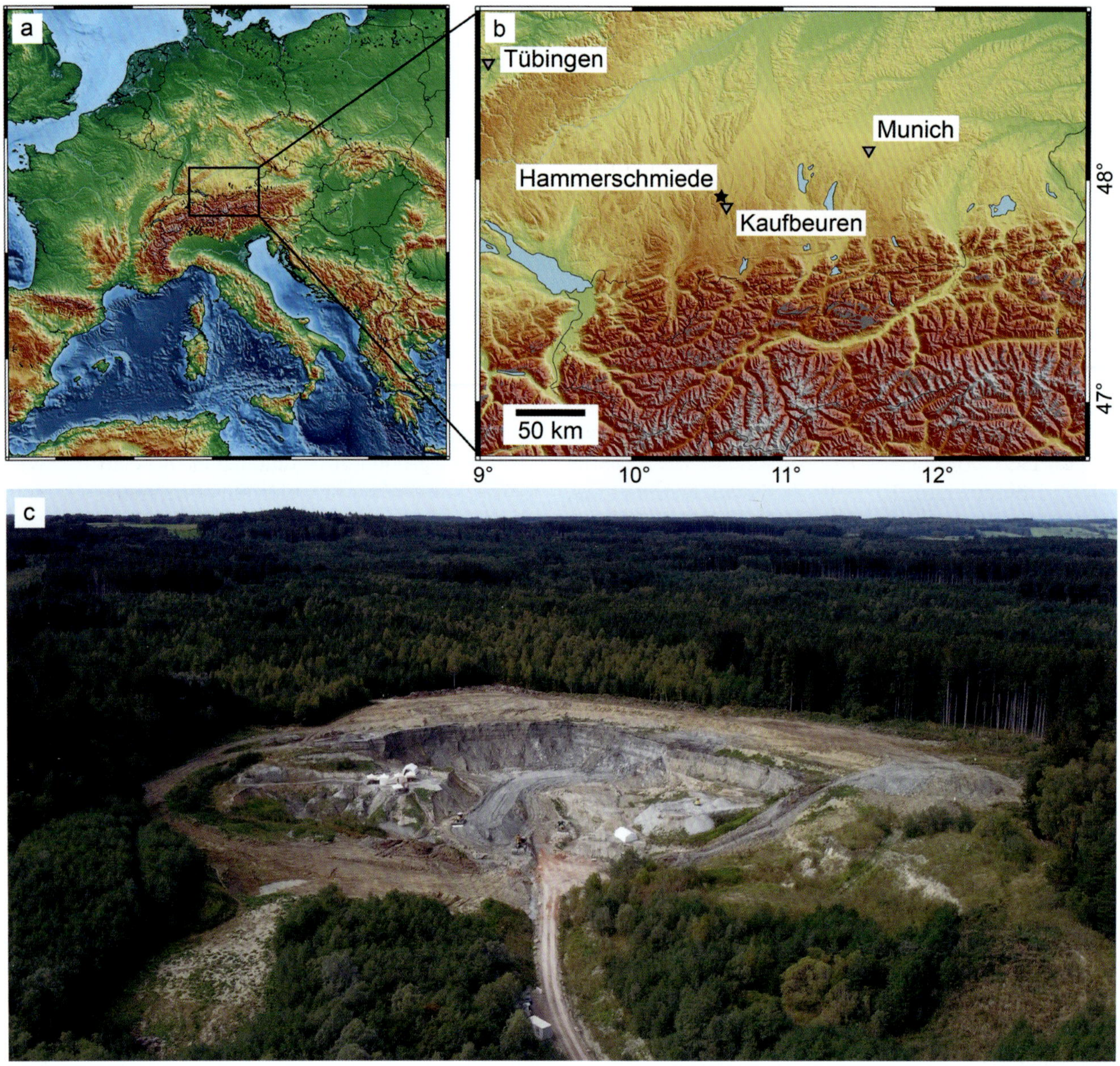

Figure 1.1.1. a) Topographical map of Europe. **b)** Enlargement of the western part of the North Alpine Foreland Basin (south German Molasse Basin). The Hammerschmiede locality is highlighted with a black star. Modified from Böhme et al. (2019a). **c)** Drone image of the Hammerschmiede clay pit looking west, taken in September 2022.

A scientific report on palaeobotanical (lignite, leaves, and palynology) finds and invertebrate fossils (helicid snails) has been provided by Meyer (1956). As early as 1965, the amateur archaeologist and private collector Sigulf Guggenmos (Dösingen) possibly made the first vertebrate discovery (turtle carapace fragment). The first discovery of vertebrates (microvertebrates) was made by Helmut Mayr (Munich) in May 1973 after S. Guggenmos had informed him of further vertebrate fossil finds around 1972. A very short description of the single fossil-bearing stratum HAM1 was reported by Mayr & Fahlbusch (1975) and Fahlbusch & Mayr (1975).

The private collectors Sigulf Guggenmos (Dösingen) and Manfred Schmid (Marktoberdorf) continued documenting fossils from Hammerschmiede during the late 1970s and early 1980s and discovered a new fossil-bearing level (HAM6), which yielded the partial skeleton of a large adult male proboscidean – *Tetralophodon longirostris* – in addition to several isolated finds of medium and small vertebrates (Konidaris et al. 2023). Probably encouraged by this discovery, H. Mayr investigated two further levels containing small mammals in 1980 (HAM2 and HAM3). Since that time, the institutionally collected material housed at the Bayerische Staatssammlung für Geologie und Paläontologie (BSPG) in Munich has been the subject of various publications. Jung & Mayr (1980) compared micromammals and palaeobotanical elements of several southern German localities and assigned the Hammerschmiede strata to the Mammal Neogene Unit MN 9. Gregor (1982) summarised the palaeobotanical findings of

Hammerschmiede and also dealt with the discovery of a new leaf-bearing level at the base of the pit by the private collector M. Schmid in 1979. He further provided an update on the large mammal finds determined by Kurt Heissig (Munich). Schleich (1981, 1984) provided the turtle finds of Hammerschmiede. As part of her unpublished PhD thesis, Seitner (1987) analysed the microflora of Hammerschmiede. Several authors supported the placement of Hammerschmiede in the Mammal Neogene Unit MN 9, as early Vallesian (Mein 1989, de Bruijn et al. 1992), or in the late Aragonian MN 7/8 (Böhme et al. 2006, 2008, Prieto et al. 2011). In numerous publications the vertebrate fauna of the layers HAM1–3 (mammals, amphibians, reptiles, and fishes) is especially mentioned (Bolliger 1999a, Daams 1999, Fejfar 1999, Hugueney 1999, Ziegler 1999, Böhme 2002, Böhme et al. 2006, Prieto 2007, Böhme et al. 2008, Prieto & Rummel 2009a, 2009b, Klembara et al. 2010, van Dam 2010, Böhme et al. 2011, Prieto et al. 2011, Prieto 2012, Böhme et al. 2012, Prieto & van Dam 2012, Böhme & Vasilyan 2014).

From the early 2000s, co-author MB began to collect geological data at the Hammerschmiede clay pit. It was precisely in the search for the youngest Molasse Basin deposits, that the outcrop of the actively mined Hammerschmiede clay pit on the western side of the Wertach River near Pforzen, a few kilometres northwest of the city of Kaufbeuren at the Allgäu region (Bavaria, Southern Germany), once again became the focus of scientific research. After a long break of research activity at this locality, the DFG funded a research project (grant BO1550/16 of the Deutsche Forschungsgemeinschaft, DFG, 2010–2013), including a borehole, that was drilled on a hill just a few kilometres to the southwest of Hammerschmiede, near the monastery Irsee (Kirscher et al. 2016).

The drill core (150.5 m) recorded 28 metres of unconformably superimposed gravel, a Pleistocene fluvial terrace (meltwater gravel of the Günz glaciation), and 122 metres of Miocene sediments of the youngest part of the Upper Series (*Obere Serie*). In contrast to the drill core, the Pleistocene cover at the Hammerschmiede clay pit is only a few metres thick and is composed of a glacial till of Middle Pleistocene age (Riss glaciation) (Kirscher et al. 2016, Fig.1). The Miocene sediments of the drill core and Hammerschmiede are mainly composed of fluvial sediments (fine sand, silt, clay) and lignite seams and mainly represent river and floodplain deposits. A magnetostratigraphic correlation of the sediments exposed at the Hammerschmiede clay pit was performed by Kirscher et al. (2016), and the 26 metres Hammerschmiede section was dated between 11.67 Ma and 11.42 Ma. Consequently, the sediments precisely cover the Middle to Late Miocene transition (late Serravallian and earliest Tortonian) (Kirscher et al. 2016). In 2011, an autochthonous mollusc fauna within a fluvial channel (HAM4, discovered in 2008) was described by Schneider & Prieto (2011). In the same year, excavations began, led by co-author MB in a smaller fossil-bearing channel, discovered by her and Jerome Prieto in 2008 in the central pit profile (HAM5), which were continued and increasingly intensified over the following years. Following initial small-scale sampling, extensive excavations have also been conducted in the local stratigraphic level HAM4 since 2017.

The excavation work has not yet been completed but more than 40,000 finds from more than 1200 square metres of excavated area have been documented so far. In a new generation of publications, the new finds from the fossil layers HAM5 and HAM4 were supplemented by finds from HAM1–3 and HAM6. Fuss et al. (2015) and Hartung et al. (2020) studied the boselaphin bovids (*Miotragocerus monacensis*, Bovidae, Artiodactyla) of the locality. Hartung & Böhme (2022) published the first evidence of cranial sexual dimorphism in tragulids (*Dorcatherium naui*, Tragulidae, Artiodactyla). Lechner & Böhme (2022, 2023) published two beaver species (*Steneofiber depereti* and *Euroxenomys minutus*, Castoridae, Rodentia) from Hammerschmiede and presented palaeoecological observations. Several publications reported on the first avian finds from Hammerschmiede, including the material of a darter (*Anhinga pannonica*, Anhingidae, Suliformes), the skull of a crane (Gruidae, Gruiformes), a leg of a new anseriform (*Allgoviachen tortonica*, Anatidae, Anseriformes), and material possibly belonging to a dabbling duck (cf. *Mioquerquedula*, Anatidae, Anserifomes) (Mayr et al. 2020a, 2020b, 2022). This was supplemented by a comprehensive description of the carnivore fauna of Hammerschmiede (Kargopoulos et al. 2021a, 2021b, 2021c, 2022, 2024a, 2024b), the chalicotheriines (Kampouridis et al. 2024) and a study of the proboscidean finds (Konidaris et al. 2023). Nevertheless, the new great ape species of *Danuvius guggenmosi* Böhme et al. (2019a), an arboreal bipedal hominid, remains the most famous discovery of Hammerschmiede and continues to attract public and scientific attention (Böhme et al. 2019b, Kivell 2019, Böhme et al. 2020, Williams et al. 2020). In 2024, another new hominid species, *Buronius manfredschmidi*, was finally added to the list of great apes, providing the first documented case of crown-hominid syntopy in the fossil record (Böhme et al. 2024).

1.2 Introduction to taphonomy

The study of fossils and fossil localities extends beyond the mere documentation of discovered specimens; it also involves understanding the processes that played a crucial role in their preservation. In principle, fossils represent an exception rather than the norm, as the vast majority of all organisms that have ever lived were not incorporated into the fossil record but were instead recycled through natural decomposition processes. The scientific discipline that investigates these processes is known as taphonomy.

The term taphonomy was first introduced by the Russian palaeontologist Iwan Antonowitsch Efremov (alternatively transcribed as Yefremov) in 1940. Efremov defined taphonomy as the totality of processes governing the transition of an organism from the biosphere to the lithosphere (Efremov 1940, Müller 1963). The term itself is derived from the Greek words táphos (τάφος), meaning "burial", and nómos (νόμος), meaning "law", collectively translating to "the laws of burial" or "the study of embedding processes".

While taphonomy is particularly significant in palaeontology – alongside taxonomy – it also plays an essential role in archaeology and zooarchaeology. Expanding on Efremov's original concept, Behrensmeyer & Kidwell (1985) refined the definition to describe taphonomy as "the study of preservation processes and their effects on the information contained within the fossil record."

Taphonomy is generally divided into three major subdisciplines (Behrensmeyer & Kidwell 1985): (1) Necrology, which focuses on the circumstances of an organism's death, the shedding of body parts, and the decay of soft tissues. (2) Biostratinomy, which examines the sedimentary processes affecting organic remains, encompassing the entire sequence from the moment of death to burial and embedding (Laatsch 1931, Müller 1951, 1963, Behrensmeyer & Kidwell 1985). (3) Diagenesis, which deals with fossilisation and lithification, particularly the chemical and physical alterations that occur post-burial.

Even after fossilisation, palaeontologists, archaeologists, and excavators contribute to taphonomic processes by exposing specimens during excavation. Post-excavation handling, storage, and curation further influence the state of fossils, meaning that these are subject to ongoing alteration. Systematic biostratinomic observations can be traced back to Johannes Walther in the late 19th century. The term biostratinomy was formally introduced by Johannes Weigelt (1919), who initially defined it as the study of fossil orientation and placement within sedimentary deposits (Weigelt 1919, 1928, Müller 1963). Today, biostratinomy has expanded in scope, now focusing on distinguishing between autochthonous and allochthonous deposits – i.e., whether an organism was embedded in situ (at the site of life or death) or transported before deposition (Müller 1963).

In both cases, external and internal forces may influence the final positioning of a carcass or skeletal remains. These forces include scavenging activity, current-induced transport, gravity-driven displacement, bloating caused by decomposition gases, and subsurface disturbances such as plant root penetration (Müller 1963). Building on this framework, Etter (1994) further refined biostratinomy into six distinct phases: (1) death and decomposition, (2) transport and sorting phenomena, (3) disarticulation, (4) fragmentation, (5) corrosion and abrasion, and finally (6) accumulation processes.

Taphonomic investigations are essential for evaluating fossils and understanding the circumstances of their deposition, particularly in fluvial-dominated deposits such as the Hammerschmiede site. These analyses provide crucial insights into the formation of deposits, enabling the reconstruction of past ecosystems. A multitude of small details must be examined as indicators of depositional processes. This includes determining which components are of autochthonous origin and which may have been transported from elsewhere (allochthonous). Additionally, distinguishing between taxa that lived synchronously and those relocated from older to younger deposits is fundamental.

Beyond establishing the absolute age of a site, it is also critical to assess the temporal resolution of individual sediment layers. Furthermore, identifying the agents responsible for the enrichment or depletion of certain fossil components allows for a more refined palaeoenvironmental interpretation. Finally, it must always be considered that only preserved components and structures are available for study and that fossilisation biases prevent all original material from entering the fossil record to the same extent.

In fluvial environments, fossil deposition and transport are influenced by differences in size, density, and surface properties (Voorhies 1969, Behrensmeyer 1975, 1988). As a result, currents and river dynamics can lead to the accumulation or depletion of fossils within certain areas (Voorhies 1969, Behrensmeyer 1975, 1988). Bone surface maturity (e. g., degree of abrasion) provides an indicator of transport distance, as bones undergo increasing abrasion and fragmentation with prolonged exposure to hydraulic processes. A skeleton entering the fluvial system as an articulated carcass may be transported either intact or in sections, depending on body size, decomposition stage (ligament integrity), and hydraulic conditions (Weigelt 1928, Müller 1963). Similarly, a transported carcass may undergo complete decomposition at an intermediate depositional site before its elements are further dispersed.

Fluvial deposits can also be reworked by shifting hydrodynamic conditions, such as meandering river channels that erode and redeposit older sediments (Voorhies 1969, Behrensmeyer 1975, 1988). Due to the inherently turbulent and spatially variable nature of fluvial deposits, a comprehensive taphonomic analysis – ideally incorporating multiple depositional settings – is necessary to achieve a realistic reconstruction of past environments.

In addition to transport processes, sedimentation primarily considers the depositional state – the moment when a component can no longer be transported and becomes incorporated into the sediment. Retention on the riverbed is influenced by bottom friction, adhesion, and the weight, density, and surface properties (e. g., contact area) of the components (Voorhies 1969, Behrensmeyer 1975). Ultimately, the flow energy must not exceed the forces anchoring the component to the substrate to ensure permanent deposition.

During the intermediate phase between transport and final deposition, objects can undergo orientation adjustments. For example, elongated elements may align with the flow direction due to an anchor effect created by an asymmetrical shape or a heavier end. Alternatively, those objects may roll perpendicular to the flow (Lyman 1994). The statistical evaluation of the orientation of elongated objects within a deposit can thus provide critical insights into the flow direction and velocity of the depositing current (Müller 1963, Lyman 1994).

One effective method for analysing such orientations is the use of rose diagrams or line-direction histograms, which illustrate the alignment and length distribution of elongated components such as bones or wood fragments (Kreuzer 1988, Lyman 1994). Following Frison & Todd (1986), these diagrams can distinguish between random and non-random orientations of long-axis elements. An asymmetry in the diagram is interpreted as a non-random, flow-influenced positioning of components (Frison & Todd 1986, Lyman 1994).

Previously little considered, research is now increasingly including taphonomic observations and ecological conclusions that are unfolding the overall picture of the early Late Miocene Hammerschmiede ecosystems (e. g., Böhme et al. 2019a, Mayr et al. 2020a, 2020b, 2022, Lechner & Böhme 2022, 2023, Konidaris et al. 2023). However, the circumstances of the fossil site itself were only dealt with superficially. There is little information on the constantly changing and optimising field methods and documentation techniques, which so far have generated an extremely extensive pool of metadata that can be used to obtain further information on fossil habitats and shed light on the genesis of the fossil site itself. The present study aims at presenting and discussing initial results of taphonomic observations alongside a methodological overview. It provides insights into documentation techniques, spatial distribution, and potential contexts of finds and sites, sedimentological findings, and biostratinomic observations. Given the extensive data available, many topics are illustrated through case studies, offering an exemplary overview of the dataset, which is expected to yield even more insights in the future.

2 Excavation and documentation methods

2.1 Excavation techniques and historical review

While excavations in the early days (2011–2015) were carried out over short periods of a few days or weeks with only a small number of participants, constant methodological and technical optimisation has led to a steady increase in the number and quality of finds. Digging is mostly done with shovels, spades, hand shovels, boning knives or spatula tools, and for delicate work scalpels and dissecting needles are used. Fortunately, the rough work is carried out by workers from the clay mining company using large construction machines to remove the several-metres-thick overlying sediments, while the final exposure of the fossil layer is accomplished independently with a small excavator.

Stable finds are removed directly from the field and washed. Brittle or fragile objects are stabilised on the site with superglues of different viscosities. Usually, the moist matrix sediments prevent the adhesive from binding the matrix too strongly with the bone. More complex finds are recovered in a plaster jacket and prepared in the laboratory. Since 2021, more challenging finds have been examined by micro-CT scanner before preparation to facilitate the subsequent procedure.

2.2 Wet sieving

In 2016, clay mining in the pit reached the areas of HAM5 that had been exposed for the first time over a large area. In a large-scale securing operation, about 23 tonnes of the fossil layer were extracted with an excavator and stored in a nearby small shed in front of the pit. At this time, there was no other way to protect the fossil-bearing strata in situ, especially to prevent the potential loss of *Danuvius* specimens (Böhme et al. 2019a). This pile of material (called "Schlämmhaufen" in the German documentation) (Fig. 2.2.1 a) was processed by screen washing since 2016 in parallel to excavations in the clay pit itself until it was completed in the 2019 excavation year and finally yielded 6 additional specimens of *Danuvius* which could thus be saved from destruction.

The first construction used for screen washing is typical for palaeontological excavations, the material was wet sieved into three grain sizes (>5 mm, >1 mm, >0.6 mm) by stacked sieves with a direct water con-

Figure 2.2.1. Overview of the first screen washing installation (August 2016) for the processing of the fossil sediments from the early Late Miocene Hammerschmiede site and the local stratigraphic level HAM5. (a) A pile of secured and stored sediments from the fossil-bearing layer HAM5 in 2016, approximately 23 tonnes. (b) The sediments were inspected for fossils, carefully coarsely crushed, and soaked with water in buckets for about 30 minutes. (c) The soaked material was screen washed and sieved into three grain sizes (>5 mm, >1 mm, >0.6 mm) using a stacked sieve screen wash construction and water from the nearby hydrant.

nection to the nearby hydrant. This work was done a few hundred metres in front of the clay pit area near a small stream into which the waste water could be discharged. At this time excavations were taking place only in HAM5. The dried clayey material from HAM5 was carefully coarsely crushed and soaked with water in buckets for about 30 minutes. At this stage, about 60 buckets of clayey mud of 10 kg each could be processed in one day (Fig. 2.2.1 b–c). The coarsest fraction had already been searched for fossils in the field. The medium grain size was transported completely and the smallest grain size in a representative sample in bags to Tübingen for further processing and picking in the laboratory (see section 2.3 Processing of the wet sieve residue).

The extreme effort to cope with the large amounts of material and finally the possibility of examining all excavation tailings for small fossils led the co-author TSL to design and construct a more optimised screening device in 2019. Based on existing designs used in industrial gold mining in North America, for example, the type selection fell on a portable sieve trommel washing system or rotational sieving system, short "Rosie" (Fig. 2.2.2). Rosie includes a horizontally rotating screen trommel with an adjustable angle, into which wet material is fed from a prewashing section. The trommel is powered by a chain drive from a windshield wiper motor, running on a 12 V car battery supported by solar panels, and rotates at a few revolutions per minute. The material is continuously washed from above by a spray bar, while the coarser fraction moves steadily toward the drum outlet due to the slight inclination. Various sieve mesh sizes are arranged along the drum sections, enabling the sorting of grain sizes and targeted removal from fine to coarse fractions. Because the particles are not thrown around as some might assume, but are slowly transported in the water film of the rotating system, hardly any damage is caused to the fossils by the washing process. In this way, more than 450 buckets per day could be processed of the clay material of HAM5 and considerably larger quantities of the sand material of HAM4 which does not need to be pre-soaked. Rosie sorts the residuals into a grain size of 1–5 mm and >5 mm. The latter is now sorted out on site in the old screen wash plant, while the former grain size is packed in bags and transported to Tübingen for further processing in the laboratory. With the completion of processing of the material from HAM5 stored outside the claypit, Rosie was directly installed and used at the excavation areas for the first time in 2020 (Fig. 2.2.2 a). The water required for this is fed to the system via around 750 metres of type B fire hose. The wastewater exits the pit via a settling basin and the site's integrated drainage system, from where it is conveyed to the nearest stream. Since 2020, all tailings have been screened for small fossils, and with the certainty of discovering all the little things, the excavation speed was increased to a higher efficiency.

Figure 2.2.2. Overview of the fossil screening zone at the early Late Miocene Hammerschmiede locality at the HAM4 excavation area in August 2023. (a1) Water supply by type B and C fire hoses over 700 metres from the next hydrant, (a2) piled up sediments, to be washed with (a3) the self-constructed rotational sieving system "Rosie", that is powered by a solar panel (a4) with a car battery as buffer storage. The old sieve rack (a5) is used for picking the washed material. The area's base is slightly inclined and covered with pond foil and drainage pipes (a6) to effectively drain the wastewater. (b) Inside view of „Rosie". A spray bar continuously cleans materials slowly transported to the outlet due to slow rotation and slight inclination of the system. (c) Technical overview of "Rosie": feed funnel with prewash and controlled feeding of the material (c1); wiper motor and a motorcycle chain as drive (c2); the sieve trommel contains two mesh sizes (1 mm and 5 mm) (c3); water supply (c4); Discharge section of the cleaned and sieved grain size 1–5 mm (c5) and >5 mm (c6).

Figure 2.3.1. Overview of the particle grain sizes portion 1–5 mm from the early Late Miocene Hammerschmiede site and the local stratigraphic level HAM4. (a) Before and (b) after processing with 10% acetic acid solution. Scale bars equal 10 mm.

2.3 Processing of the wet sieve residues

In HAM4, about 10% of the wet sieved volume of tailings remains as residue. 20% of this corresponds to the coarse fraction (>5 mm) and 80% to the size range 1–5 mm. The latter fraction contains about 1% phosphatic (bones and teeth), siliciclastic, carbonate (shell fragments), and organic particles. The remaining part consists of approximately equal proportions of moist clay pebbles (resistant to first wet sieving) and pedogenic carbonate concretions. After drying and subsequent soaking in water, the clay portion disintegrates and can be washed out, and the remaining residue can be directly picked for microfossils to document carbonate fossils (Fig. 2.3.1 a). The larger portion of the residue is further concentrated through the addition of acid.

The use of diluted acetic acid 5–10% has proven to be particularly gentle on bones and enamel so that all carbonate components (shells and pedogenic carbonates) are dissolved, and only phosphatic, siliciclastic, and organic components remain (Fig. 2.3.1 b). This procedure provides a minimal volume with maximum accumulation of osteological material for the microscopic search for microfossils, as only 1% of the material collected in the field and even only 0.1% of the total excavated sediment remains as residue here. Bones and teeth are not damaged by this process. The remaining material consists of siliceous, organic, and phosphatic (bone) grains and is subsequently picked for microfossils under the binocular microscope. The processing of larger amounts of sieve residues began in 2016 (HAM5) and 2020 (HAM4) and has been intensified since 2023. Although still ongoing, the work has already yielded promising results, with substantial material yet to be processed and further finds expected.

2.4 Documentation of finds

2.4.1 Spatial documentation

Since the 2017 excavation, osteological specimens from a size of about 2 cm but also particularly interesting smaller finds are documented on-site regarding their spatial location and distribution. For this purpose, total stations of the type Leica TCR 407 power (used during excavation period 2017–2018) and geo-FENNEL FTS 102 (from 2019 onward) are employed. The stationing of the total stations is done using fixed points in the pit area and the excavation areas. Long objects are documented with at least 2 measuring points to obtain information on length and orientation. More complex situations, such as partly articulated or arranged finds, were documented photographically and/or by find sketches.

2.4.2 On-site registration of specimens

Since the excavation year 2015, an excavation register has been maintained, providing preliminary object identification and a chronological sequence of finds. Specimens from a size of approx. 2 cm and, in the case of more important finds, also below this size are registered with an identification number at the site. For

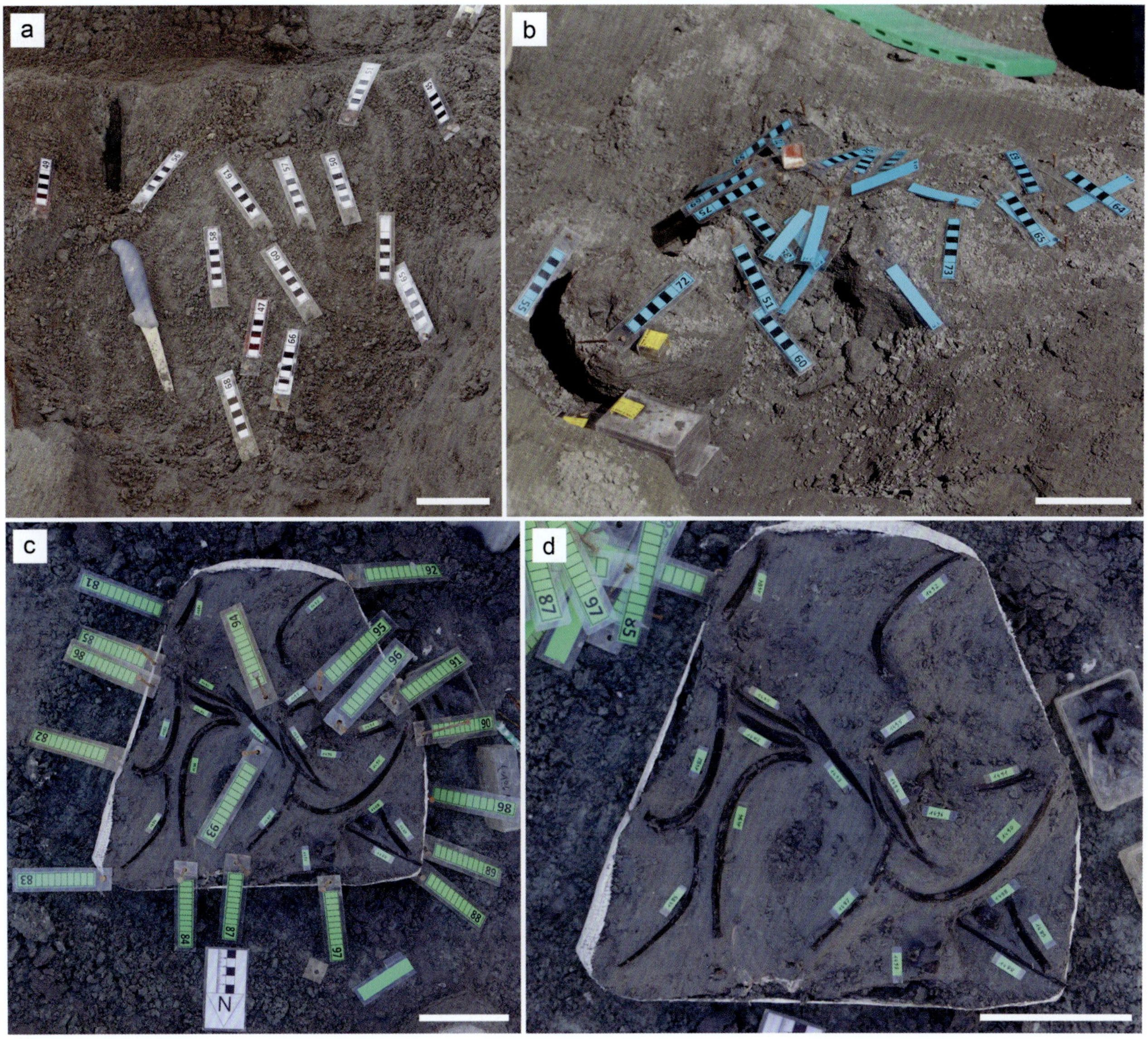

Figure 2.4.2.1. On-site views of the documentation of isolated finds (a) and find complexes (b–d) at the early Late Miocene Hammerschmiede site and the local stratigraphic level HAM4. (a) Particularly specimen-rich spots with finds deposited incoherently next to each other (August 2021). (b) Number tags mark multiple juvenile tragulid (*Dorcaterium naui*) bones, which presumably resemble a single individual (August 2022). (c–d) Complex assemblage of ribs and vertebrae of an adult tragulid (*Dorcaterium naui*) stabilised by a plaster jacket (August 2019). Note that (a) does not contain any objects but only the markers used for measurements, and (b) does contain the objects, although they are not visible as they have not been prepared or cleaned in situ. The number tag colours correspond to the following ranges: green 14XX; white 37XX; blue 89XX. Scale bars equal 10 cm.

surveying and documentation purposes, the item is removed, and its position in the field is marked with a number tag and steel nail until spatial measurements are completed (Fig. 2.4.2.1).

Due to rainwater accumulating on the excavating surface, laminated number tags are used to mark the findspots, which correspond to a certain number range due to their color at the respective point in time (Fig. 2.4.2.1). The advantage of this method is that the flags can be reprocessed, sorted, and reused quite quickly. Longer objects are marked with a second number tag (handwritten A- and B-number) to generate two measurements for directional analysis (Fig. 2.4.2.1 b). Find complexes not recovered individually in the field were stabilised on site with a plaster jacket. A small handwritten label with the number is placed next to every visible registered bone and plastered together with it, so that an assignment to the inventory number remains until the final preparation in the laboratory (Fig. 2.4.2.1 c).

Since 2017, all finds marked with find numbers can be reconstructed with a spatial survey (X, Y, Z coordinates) with an accuracy of up to 1 mm. For finds from the wet-sieved tailings, a spatial reconstruction of the origin has been dispensed with so far. However, the tailings can be documented to the day and at least assigned to the corresponding excavated sectors. In this way, overlooked fragments can be more easily assigned and refitted to surveyed specimens, and potential find connections (strewnfields) can be verified or rejected.

2.5 Excavation crew

In the early years (2011–2016), the excavation participants consisted mainly of staff and students of the University of Tübingen working group. In the following years, the participation of volunteers from the surrounding area of the fossil site, as well as from Germany and abroad, increased. In the meantime, more than 300 people are registered on our participant lists, and in the last season (2024), 114 volunteer participants were able to take part in our citizen science excavation project – "Bürgergrabung Hammerschmiede" – while in the same year, the employed crew for the organization consisted of only eight people.

2.6 Sample deposition

The samples and finds from the local stratigraphic levels HAM1, HAM2, and HAM3 from earlier field work at the Hammerschmiede clay pit made by Helmut Mayr are stored in the Bavarian State Collection of Palaeontology and Geology in Munich, Germany and labelled as Hammerschmiede 1 (HAM1): SNSB-BSPG 1973 XIX, Hammerschmiede 2 (HAM2): SNSB-BSPG 1980 XXVII, and Hammerschmiede 3 (HAM3): SNSB-BSPG 1980 XXVIII. The material from HAM4 and HAM5 has been collected during excavations conducted by the University of Tübingen, Germany, since 2011. All material collected since then is stored in the Palaeontological Collection of the University of Tübingen, Germany (GPIT) and is labelled with numbers of both GPIT – for excavations from 2011 until 2019 – and SNSB-BSPG, for the excavations in the period 2020 to 2024 where the latter are indicated as Hammerschmiede 4 (HAM4): SNSB-BSPG 2020 XCIV and Hammerschmiede 5 (HAM5): SNSB-BSPG 2020 XCV. The private collection of Sigulf Guggenmos (Dösingen) from the find layer Hammerschmiede 6 (HAM6) was donated to the Palaeontological Collection of the University of Tübingen (GPIT), Germany in 2018. A few finds from Hammerschmiede are stored in the collection of the Kempten-Museum im Zumsteinhaus, Kempten, Germany, and in the private collection of Manfred Schmid (Marktoberdorf) with casts of more relevant objects stored in Tübingen (GPIT).

2.7 Quantity of material

The constantly optimised techniques and methods allow for ever more effective excavations, which are constantly reflected in the growing number of specimens per field season (Tab. 1). The find numbers are

Table 1. Quantity of material excavated at the Hammerschmiede fossil site and the corresponding data for local stratigraphic levels HAM5 and HAM4 respectively. Note, that geospatial measurement of objects started in the 2017 field season. The specimens recorded until then (marked with an asterisk) resemble the field book inventory. It should also be noted that an extremely large number of microfossils are only archived in collection boxes or are still waiting for further processing so the actual number of finds is significantly higher. The isolated finds of HAM4 for 2023 and 2024 are still being processed, but a quantity of about 3500 objects is expected.

HAM4	**2017**	**2018**	**2019**	**2020**	**2021**	**2022 South**	**2022 North**	**2023 North**	**2024 North**	**TOTAL**
Area [m²]	56	70	197	60	70	140	50	85	132	**860**
Measured nr.	534	528	3,164	1,089	1,343	1,323	160	669	680	**9490**
Isolated finds	539	224	719	2,345	2,729	3,811	0	2,874	~1,200	**~1,4441**
Total	**1,073**	**752**	**3,883**	**3,434**	**4,072**	**5,134**	**160**	**3,543**	**~1,900**	**~2,3951**

HAM5	**2014**	**2015**	**2016**	**2017**	**2018**	**2019**	**2020**	**2021**	**2022**	**2023**	**2024**	**TOTAL**
Area [m²]	2	10	25	40	100	0	50	100	2	0	37	**366**
Measured nr.	139*	368*	711*	342	833	0	260	215	23	0	48	**2,939**
Isolated finds	158	81	104	136	50	0	80	1	0	0	~150	**760**
Total	**297**	**449**	**815**	**478**	**883**	**0**	**340**	**216**	**23**	**0**	**~198**	**3,699**

subdivided into measured objects, those that were already discovered in situ and spatially documented on site, and isolated finds ("Lesefunde", Rosie specimens), those that were only registered in the wet sieving or from the collection boxes of the individual excavators afterwards. The numbers of the corresponding excavation years can be taken from Table 1. Note that geospatial measurements of finds have only taken place from the 2017 field season onwards. All numbers from previous years correspond to the numbers registered in the field book. It should also be noted that an extremely large number of microfossils are only archived in collection boxes or are still waiting for further processing, so that the actual number of finds is significantly higher.

3 Geology and stratigraphy of the clay pit Hammerschmiede

The actively mined Hammerschmiede clay pit exposes up to 27 metres of Miocene fluvio-alluvial sediments, supplemented by two major lignite seams – one near the top and one close to the base of the profile. Alluvial deposits include clays, silty clays, sands, marls, and silty marlstones. The sedimentary succession and its stratigraphic correlation in the north–south direction have been described by Kirscher et al. (2016). Due to the progression of the working area in the western direction during the last 10 years, the succession can now be better understood three-dimensionally

Figure 3.0.1. Overview of the Hammerschmiede clay pit, showing the approximate positions of the local stratigraphic levels HAM5 and HAM4 (February 2021). (a) Drone image from the east, in the direction of active clay mining. (b) Indicated areas of the local stratigraphic levels HAM5 (blue) and HAM4 (yellow). "HAM4 South" and "HAM4 North" correspond to the same channel HAM4, which artificially was interrupted in the central pit area due to mining activities. Image width approximately 150 metres.

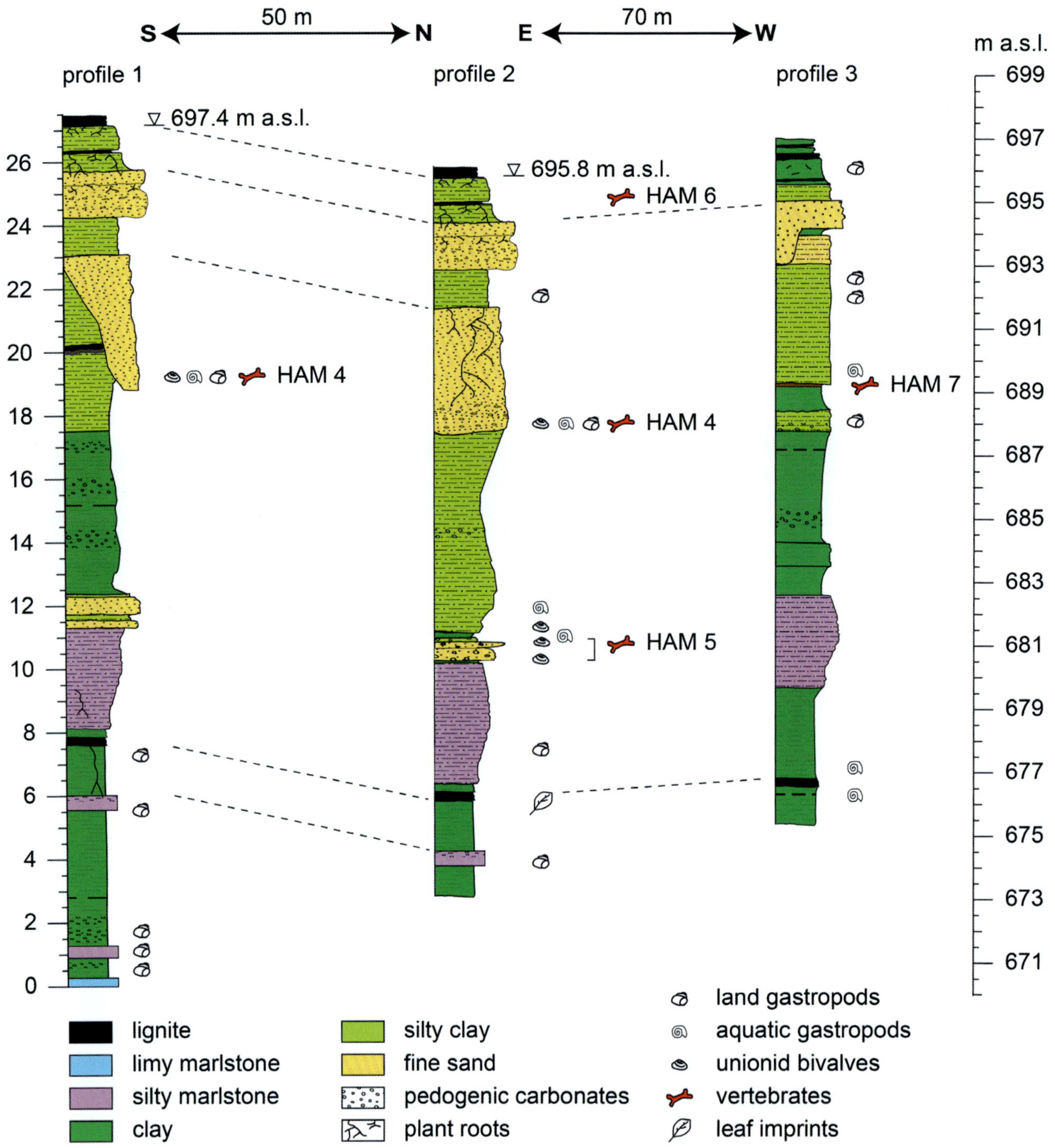

Figure 3.0.2. Lithological logs of three sections from the early Late Miocene Hammerschmiede section. Profile 1 marks the southernmost section of the clay pit. Profile 2 is located approximately 50 metres to the north of profile 1. Both profiles were exposed from 2011 till 2015 and are described in Kirscher et al. (2016), but are now lost due to clay mining. Profile 3 lies 70 metres to the west of the second profile and has been exposed since 2021. Fossil-bearing layers HAM4, HAM5, and HAM7 are marked in the profiles, and the supposed location of the historical find layer HAM6 from the end of the 1970s is noted. Note that the colour codes do not correspond to Fig. 3.0.3. The figure is updated and supplemented based on Kirscher et al. (2016: fig. 1c).

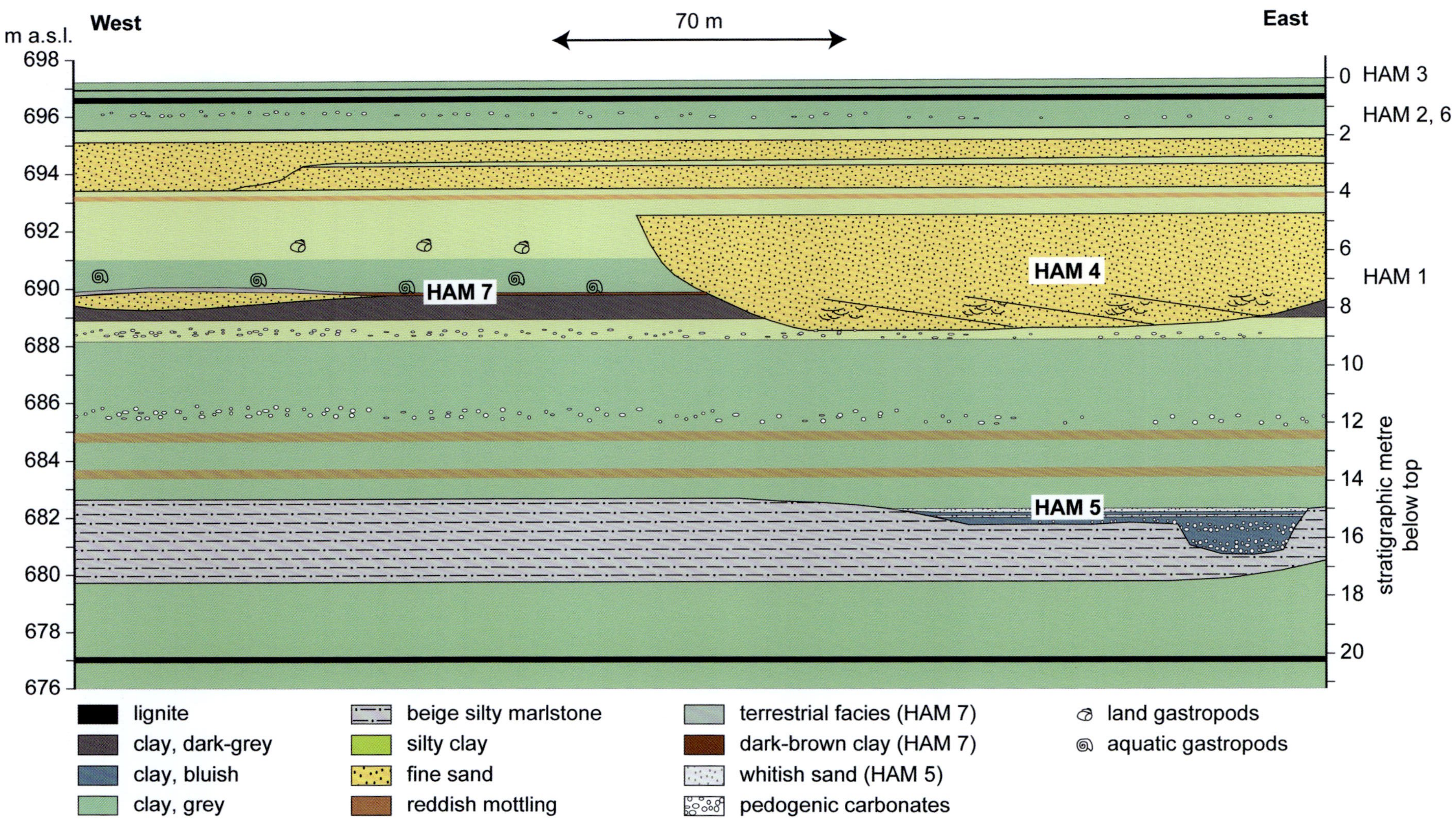

Figure 3.0.3. Detailed stratigraphic west-east section of the early Late Miocene Hammerschmiede (HAM) locality with lithological determinations. The fossil fluvial channel deposits of HAM5 and HAM4 are marked within the section, and the recently discovered HAM7 is highlighted. Picture width corresponds to 70 metres in the central pit area with a view to the north. The elevation above sea level refers to the centre of the clay pit.

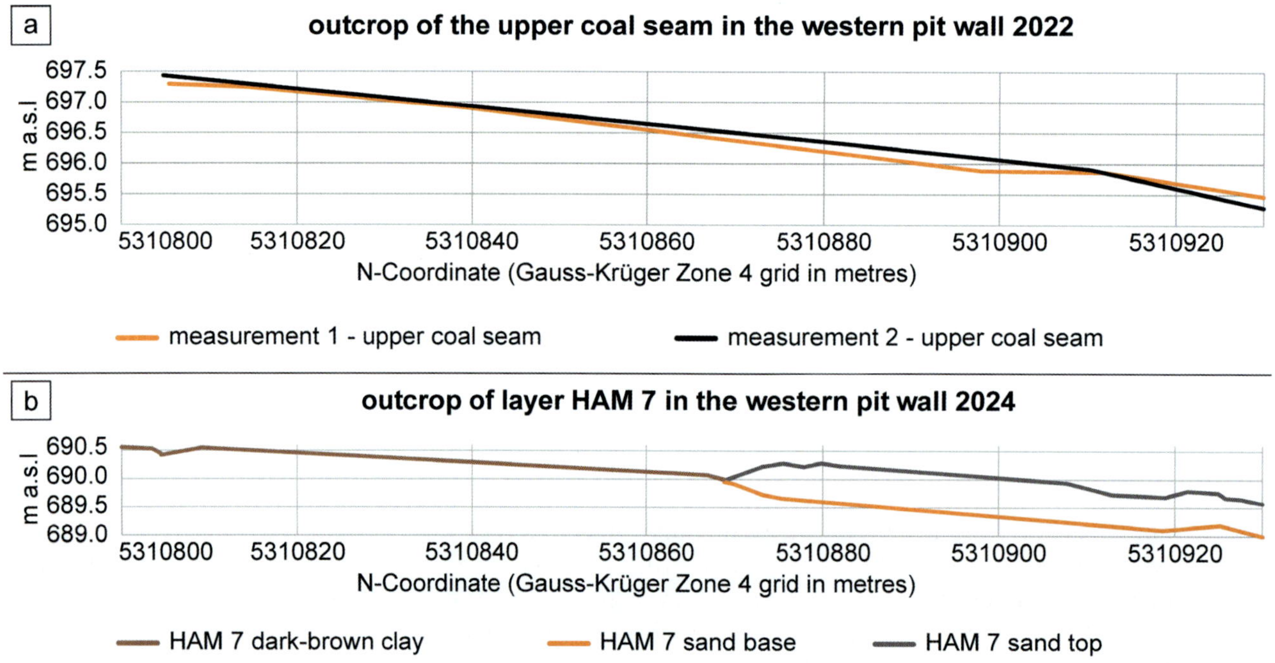

Figure 3.0.4. Different marker horizons and their elevation progression in the western outcrop of the Hammerschmiede clay pit. (a) The upper coal seam with two measurements from different viewpoints in 2022. (b) The new find layer HAM7 in the western part of the pit wall (2024), lies somewhat deeper in the profile. The stratum begins with a dark-brown layer, which laterally contacts a lense-shaped crevasse-splay sand. The dark-brown layer and the sand body are measured separately. Coordinates correspond to Gauss-Krüger Zone 4 grid in metres.

(Figs 3.0.1, 3.0.2). The west–east profile (Fig. 3.0.3.) reveals that the Hammerschmiede succession can be subdivided into two parts. The lower sequence between 671 and 690 m a.s.l. is dominated by clay and silty clays, which show variable pedogenic overprint expressed by lessivated soil horizons, rich in pedogenic carbonate concretions or reddish mottling. The mollusc fauna is entirely terrestrial (Kirscher et al. 2016). This sequence encompasses the lower coal seam at 677 m a.s.l. and the up to three metres thick silty marlstone of yet enigmatic origin. The only evidence for erosional processes is the HAM5 channel cut into the silty marlstone (see below). In contrast, the upper sequence, between 690 and 695 m a.s.l., is dominated by erosive fluvial processes. The sand-dominated facies are constituted by channels of meandering rivers (HAM4 channel, see below), corresponding crevasse splays (HAM7, see below), and lacustrine clays rich in *Bithynia* mud snails (between 690 and 691 m a.s.l.). The uppermost two metres of the present outcrop may indicate the recurrence of a clay-rich sequence, as suggested by palaeosoils with pedogenic carbonates and a major (and two minor) coal seam.

The sediments of Hammerschmiede dip northward. Total station measurements were used to determine the tilt of the sediments. The curved quarry edge of the western pit provides the clearest and most continuous profiles of the clay pit. Along an outcrop line of more than 130 metres, it was thereby possible to observe the continual course of different marker horizons. The blackish upper lignite seam and the dark-brown new find layer HAM7 were selected as marker horizons for this approach. These layers were easily recognizable from a distance and are consistent in their thickness. Measurements were taken using a total station with a viewpoint from the corresponding excavation areas of HAM4 South and HAM4 North, providing the best fixed-point stationing coverage. Due to potential measurement errors at these large distances (positional inaccuracies), the measurement for the upper coal seam was repeated from different viewpoints (Fig. 3.0.4 a). Layer HAM7, however, could only be surveyed and measured from the excavation area HAM4 North, which meant that only one set of measurements was taken (Fig. 3.0.4 b). The lateral change of sedimentary facies in HAM7 makes it necessary to consider this layer in two parts. All observed layers dip northward (Fig. 3.0.4 a–b). The upper coal seam dips at approximately 0.85° (1.5%) to the north (Fig. 3.0.4 a). The dark-brown HAM7 shows a dip angle of 0.5° (0.8%), although the HAM7 crevasse sand base (northern HAM7 transition) dips at 0.7° (1.2%) (Fig. 3.0.4 b). As a consequence of the slightly different results, the layers incline by more than 0.5° in a northerly direction but do not exceed a gradient of 1°.

The measured northern dip of the layers proves that the *Obere Serie* sediments at Hammerschmiede belong structurally to the southern wing of the North Alpine Foreland Basin, the so-called Inclined Foreland Molasse (*Aufgerichtete Vorlandmolasse*, see Fig. 1.0.1;

Lemcke 1988, Doppler et al. 2005, Scholz 2016). The tilt to the north was initiated by overthrusting of advancing Alpine nappes and the resulting folding and uplift of the Folded Molasse (= Subalpine or Imbricated Molasse), located more than 20 km south of Hammerschmiede.

3.1 Fossiliferous horizons of Hammerschmiede

The numbering of the local fossiliferous layers corresponds to their chronological order of discovery, not their stratigraphy (Figs 3.0.2, 3.0.3). The local stratigraphic levels HAM1, HAM2, and HAM3 are known from previous collections at Hammerschmiede (Mayr & Fahlbusch 1975, Fahlbusch & Mayr 1975, Fahlbusch 1975, Prieto & Rummel 2009, Prieto 2012), but only HAM1 was described by Mayr & Fahlbusch (1975). This site was found by Helmut Mayr (Munich, Germany) in May 1973 in the northern part of the pit at that time (eastern edge of the present-day pit). The fossiliferous horizon was a 0.5 m thick greenish-grey marl, rich in freshwater gastropods (*Bithynia*). According to our yearlong knowledge of the outcrop conditions, such a layer occurred only once in the succession between 690 and 691 m a.s.l. (Fig. 3.0.3). The unpublished horizons HAM2 and HAM3 have been found by Helmut Mayr in 1980. According to his statements to the co-author MB in the year 2000, the HAM2 level was situated five metres above HAM1 and a major sand package. Hence, we correlate HAM2 directly above the last crevasse splay sand between 695 and 696 m a.s.l. The HAM3 level was identified by Helmut Mayr above HAM2 on top of the succession. According to the discoverer's recollection, it represents a "sandy-marly channel filling with coal fragments, around 10 m above HAM1". A fluvial environment of this horizon seems indeed possible, given the freshwater molluscs (*Margaritifera*, *Spherium*, *Pisidium*, *Borysthenia*, *Segmentina*, *Gyraulus*, two species of *Bithynia*) described or mentioned by Schneider & Prieto (2011). However, sandy channel fills with coal fragments do not exist in the present-day Hammerschmiede section. A possible solution would be that HAM3 correlates with a now-removed south–north directed channel, which cut and eroded the upper seam eastwards from the present-day outcrop. In this case, the topographic correlation of the HAM3 level would be around the level of 697 m a.s.l., but stratigraphically younger than the upper lignite seam.

The fossiliferous horizons HAM4 and HAM5 were discovered consecutively by the co-author MB together with Jerome Prieto (Munich, Germany) in 2008. During the late 1970s and early 1980s, the private collectors Sigulf Guggenmos (Dösingen, Germany) and Manfred Schmid (Marktoberdorf, Germany) discovered another spatially delimited fossil horizon (Gregor 1982), classified after 2008 as HAM6. Based on photographs, notes, and memories made by both private collectors, it was possible to identify the stratigraphic position of this site, which is no longer accessible today. HAM6 was positioned at the top of the profile at the time of discovery, below the presently exposed upper coal seam above the channel HAM4 and below glacial till (S. Guggenmos pers. comm. 2017 and 2018, M. Schmid pers. comm. 2022; see photo in Konidaris et al. 2023: fig. 3). According to a profile drawn by Sigulf Guggenmos in the early 1980ies the about one-metre thick grey clay containing the partial skeleton of *Tetralophodon longirostris* and other mammals, was positioned directly above a 20 cm thick grey mica-rich fine sand and below about one metre of brownish to greyish clay. This information matches with a stratigraphic position directly above the highest (mica-rich) crevasse-splay sheet, between 695–696 m a.s.l.. Thus, the stratigraphic positions of HAM2 and HAM6 are identical.

In August 2019 another fossil stratum – HAM7 – was discovered by the two authors of this study. This horizon emerged westward of the erosional channel HAM4 with the progression of mining. HAM7 can be traced in the east–west direction for about 40 metres (Fig. 3.0.3) and is represented by two facies: 1) a 10 cm thick dark-brown organic-rich layer with a high abundance of disarticulate fish bones and teeth in addition to finds of small beavers (*Euroxenomys minutus*), snapping turtles (*Chelydropsis* sp.), charophyte gyrogonites, and molluscs belonging to the genera *Planorbarius*, *Planorbis*, and *Bithynia*. This brownish lacustrine facies is overlying dark-grey clays with sharp contact (Fig. 3.0.3). The 2) facies is found as a direct west- and northward continuation of the lacustrine dark-brown clays. It represents a 10 cm thick layer of mud containing very abundant clasts of clay, autochthonous pedogenic carbonates, and fossils characterizing non-aquatic environments (rhino teeth and *Testudo* carapace fragments). This terrestrial facies of HAM7 is developed only above mica-rich sands of a crevasse-splay tongue. This lens-shaped sand body is at maximum 60 cm thick and pinches out rapidly after 20 metres. So, the HAM7 horizon is related to a crevasse event. The damming of the sand body created a lake lateral to it (HAM7 lacustrine facies) and established terrestrial conditions above the groundwater table on top of the sand body (HAM7 terrestrial facies). Both facies underly the lacustrine clays correlated with HAM1.

Figure 3.2.1. View of the northern wall of the Hammerschmiede clay pit in January 2020 (a) and with a sketch overlay (b). The stratigraphic hiatus between the ochre-coloured glacial till deposits of Pleistocene age at the top of the profile and the underlying early late Miocene deposits is clearly visible. The Miocene sediments are characterised by clays, silty clays, and sands (grey, green to bluish green), and one of the lignite seams (black) can be observed at the top of the pictured profile. Brighter colours indicate clay-rich strata, which dry out more quickly on the surface than the darker, sandier, and partly water-bearing levels. The erosive channel base structure represents the level and excavation area HAM4 North. HAM5 is located at the lower right edge at the base of the pit wall. Reconstructed channel width (b) as an approximate scale corresponds to about 50 metres.

3.2 The channel HAM4

The transverse section of the channel structure HAM4 is about 50 m wide and 4 m thick (Fig. 3.2.1). It preserves the main sedimentary features characterizing laterally accreting meandering rivers (Allen 1965): channel lag, lower and upper point bar deposits, and channel-fill (Fig. 3.2.2). The up to 3 m thick upper point bar and channel-fill sediments are non-fossiliferous and are removed before excavation.

For the investigation of the HAM4 channel internal mesoscale structure, two exemplary profile sections were chosen in the largest contiguously excavated area HAM4 South (2019–2023) (location of the profiles in the excavation area, see Fig. 3.2.3). Profiles were compiled transversely (Profile P1, Fig. 3.2.4) and parallel to the palaeo-flow direction approximately oriented from south to north (Profile P2, Fig. 3.2.5) of the HAM4 channel, using objects larger than 2 cm across a 4 m wide accumulation zone.

Medium and larger-sized vertebrate fossils (>2 cm) are concentrated in up to four inclined packages, interpreted as laterally accreting lower point bar units and associated channel lag deposits. Inclination occurs along the palaeo-flow direction with dip angles 2–3° (Profile P2, Fig. 3.2.5) and perpendicular to the palaeo-flow direction with dip angles 4–10° (Profile P1, Fig. 3.2.4), whereby dip angles increase with bar migration. This pattern indicates both downstream accretion

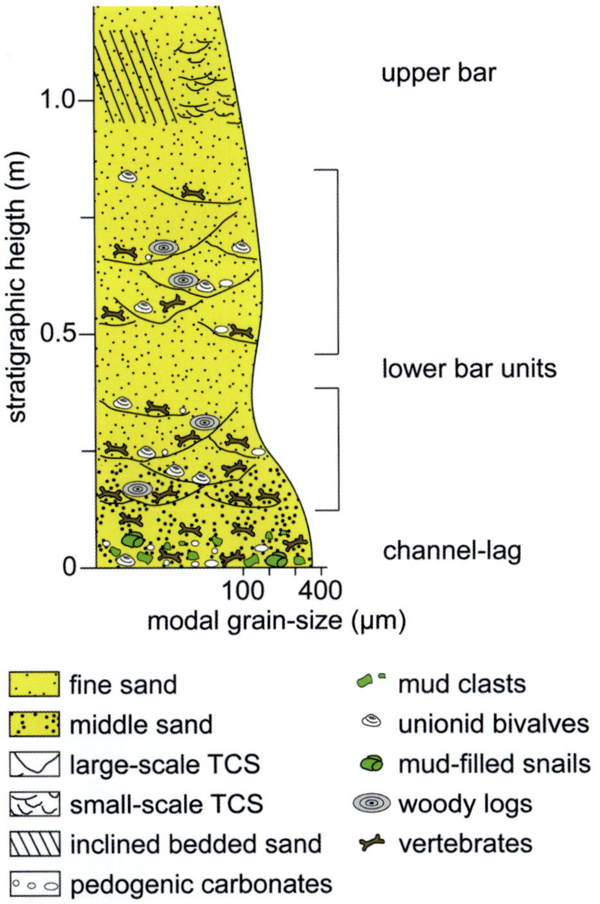

Figure 3.2.2. Lithologic log, grain size, sedimentary structures, and typical fossils and clast components in an exemplified excavation site in the channel HAM4 (local stratigraphic level at the early Late Miocene Hammerschmiede locality) showing basal channel lag deposits underlying two superimposed lower bar units and upper bar deposits (note that before excavation we remove the majority of upper bar sediments).

in northern direction and lateral accretion in eastward direction of the point bars.

The channel base consists of greenish silty clay. It shows only minor differences in surface roughness of the river bottom. In the marginal area, it is somewhat wavy and sometimes shows small depressions, while the central channel bottom of the HAM4 is a fairly flat surface at 688.5 m a.s.l. On top of the channel base, the fossiliferous packages represent lag deposits (Fig. 3.2.6). Besides vertebrate bones, mammal teeth, and freshwater pearl mussels, they contain brecciated, few cm-sized, brownish or greenish mud clasts, land gastropods filled with similar greenish mud, and pedogenic carbonate concretions, all deriving from cut-bank erosion. The modal grain size is middle sand (360 µm) and a subordinate member is fine sand (180 µm; Fig. 3.2.7). Two additional grain-size members are in the clay and fine-silt fraction. These fine components result from disintegrated mud clasts during sample preparation (carbonate leaching). Coarse sand (0.5–1 mm) has only a minor contribution to the channel lag, whereas very coarse sand (1–2 mm) is lacking and can only be picked out in screen-washed residue in very low quantities. Larger grains, up to coarse-sized pebbles (25 mm) are represented only by reworked pedogenic carbonates (Fig. 3.2.8.). Only three medium- to coarse-sized quartz pebbles have been excavated from several hundred cubic metres of channel lag. These middle sand dominated channel lag deposits are the coarsest Miocene sediments at Hammerschmiede and represent the position of the river's thalweg at a given time.

Above and laterally the channel lag follows large-scale (a few decimetres to about one metre in cross-section) trough cross-stratification (Figs 3.2.6 and 3.2.9). The trough axis is parallel to the reconstructed flow direction, thus representing the lower bar deposits. The modal grain size within a trough is in the fine sand fraction (180 µm, 58 vol%). Clay is very rare (2.1 vol%) and silt (14.6 vol%) and medium to coarse sands (25.3 vol%) are subordinate (Fig. 3.2.9).

Thus, the grain size spectrum from large-scale cross-stratified sands differs from channel lag deposits in the absence of both the middle sand mode and the mud clasts. The up to 0.5 m thick large-scale cross-stratified sands contain, next to the vertebrate fossils, commonly reworked pedogenic carbonate concretions, unionid shells, and large elongate wood fragments (Fig. 3.2.2). They show furthermore reddish streaks (schlieren) of garnet (and other heavy minerals) accumulations (compare Fig. 3.2.6). Outside a trough and towards the top of the lower bar unit fossils are very rare, but grain size distribution is similar. However, the modal size is shifted towards finer fine sand (140 µm). Silt (25.4 vol%) and fine sand (59.9 vol%) dominate over clay (3.4 vol%) and middle sand (11.2 vol%). Hence, the troughs within the lower bar deposits collect the coarser fine sand and more middle sand fraction, pebble- to cobble-sized fossils, and sunken wood fragments, all accumulated as bedload by saltation and traction.

The transition from lower to upper bar deposits is marked by the gradation into small-scale (about 10 cm in cross-section) trough cross-stratification and inclined bedded sand (Fig. 3.2.9). These very fine-sandy deposits are up to three metres thick and are mostly deposited during times of higher discharge. Unionid shells are absent here, and larger vertebrate fossils are rare. If vertebrates occur they are well-preserved and rather complete, as evidenced by an antlered skull of *Euprox* or an associated skeleton with the skull of *Chelydropsis*.

The very fine-sandy channel-fill sediments are not studied in detail here. They bear no fossils, but occasionally they preserve tree rooting systems, extending down to the base of the channel (see Fig. 3.0.2 and photo in Kirscher et al. 2016: fig. 3C).

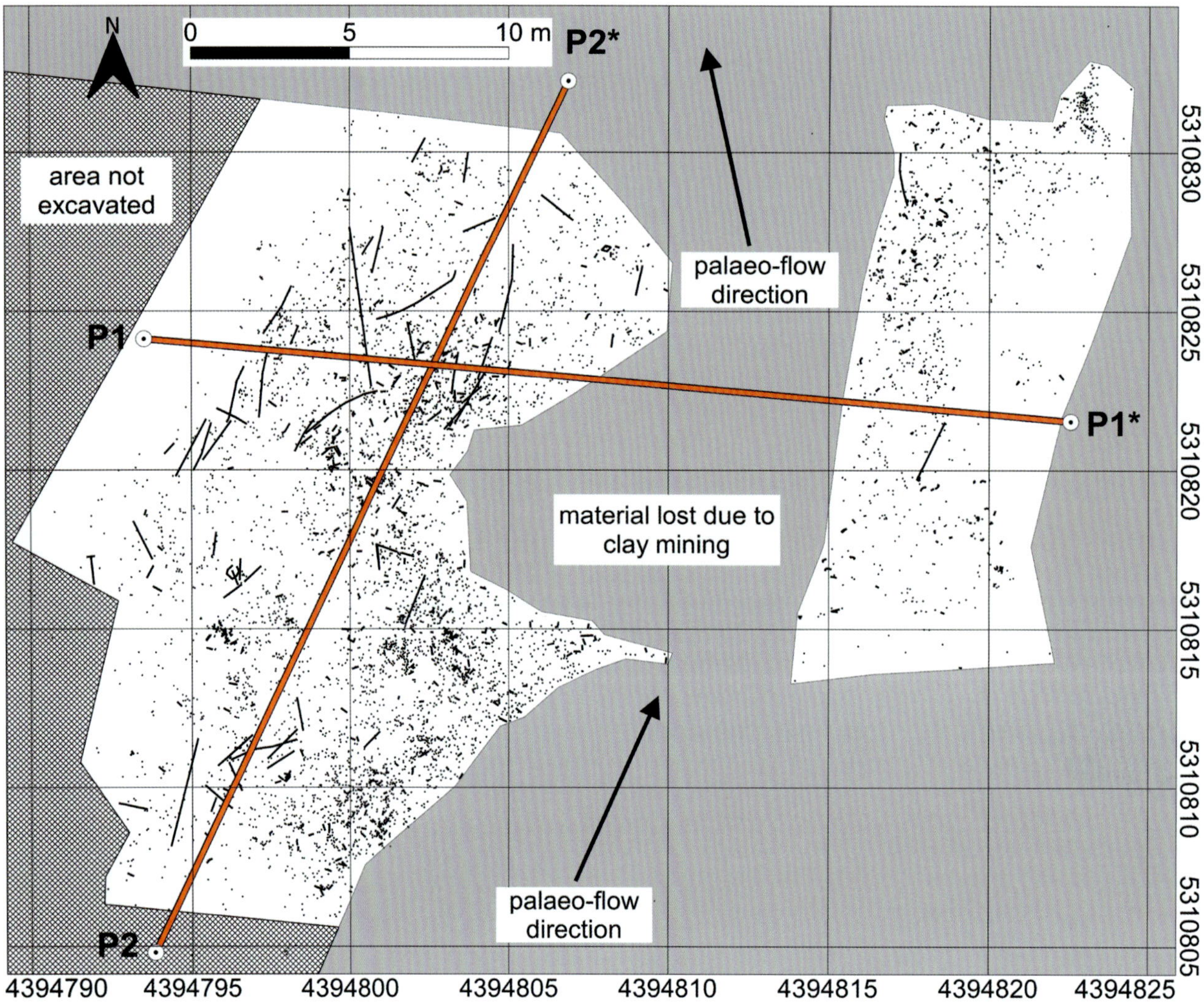

Figure 3.2.3. Detailed map of find distributions in the central excavation areas and line of the specimen accumulated profiles in the excavation area of the local stratigraphic level HAM4 (excavations 2019–2023) at the early Late Miocene Hammerschmiede locality. Data were generated during excavations from 2019 to 2023. Profiles were accumulated transversely (Profile P1, see Fig. 3.2.4) and along the palaeo-flow direction (Profile P2, see Fig. 3.2.5) of the HAM4 channel from objects larger than 2 cm over an accumulation width of 4 metres. Profile 1 median profile course runs from P1 (E4394793.52, N5310824.13) to P1* (E4394822.56, N5310821.53) and for Profile 2 from P2 (E4394793.83, N5310804.81) to P2* (E4394806.88, N5310832.25). For further detailed descriptions of this map, see section 5.1: Area of excavation and documentation, Fig. 5.1.2.2. Easting coordinates correspond to Gauss-Krüger Zone 4 grid in metres and elevation in m a.s.l.

Based on the thickness and lateral extent of the channel lag and point bar sands, we estimate the active HAM4 river to have been approximately 15 m wide and 1 m deep during mean low water level (Fig. 3.2.10). At high discharge, the bankfull depth reached 4 m, and the river width at this stage extended to 50 m toward the end of its development.

The cut-bank erosion of the HAM4 river includes sedimentary horizons accommodating the fossiliferous levels HAM1 and HAM7 (Figs 3.0.3, 3.2.1, and 3.2.10). Reworking of fossils into the HAM4 channel lag is therefore possible. However, this would not affect the biostratigraphic homogeneity of the HAM4 sample, because the levels HAM1 and HAM7 are only 10 kyrs older (see section 3.4 below), but potentially the ecological homogeneity, since the layer HAM1 and the dark-brown marl HAM7 are lacustrine in origin (see section 3.1).

3.3 The channel HAM5

The HAM5 channel-fill shows at least three successive events, beginning with a basal channel lag deposit followed up by a fining-upward sequence of silty fine sand and clay (Fuss et al. 2015). The next overlying phase begins again with a channel lag deposit that accumulated coarse-grained material as an erosive enrichment phase and ends like the first sequence described. In the fine-grained top of those sequences, root traces indicate intermediate vegetation phases and consequently a temporal separation. Both channel lag deposits consist of massive accumulations, often component-supported, of reworked pedogenic carbonates within a bluish clay matrix (see Kirscher et al. 2016: fig. 3D). The first two phases largely fill the deeper main channel and are interpreted by Böhme et al. (2019a) as "a riffle pool sequence of a small and shallow meandering rivulet with a thalweg width of 4–5 m and a maximum pool depth of 1 m". The third and last phase fills the main channel, in addition to a shallower lateral channel expansion. This last phase is also a fossil-bearing fining upwards sequence but with a lesser abundance of reworked pedogenic carbonates. It shows, especially in the area of the lateral side channel, a particular frequency of strewnfields of single individuals of many different vertebrate species. The base of the lateral channel expansion is very irregular, representing a scoured erosion surface with repeatedly elevated and hollowed areas. The often basin- or pool-like hollows and undulations of the former rivulet bottom occasionally show accumulations of fossil

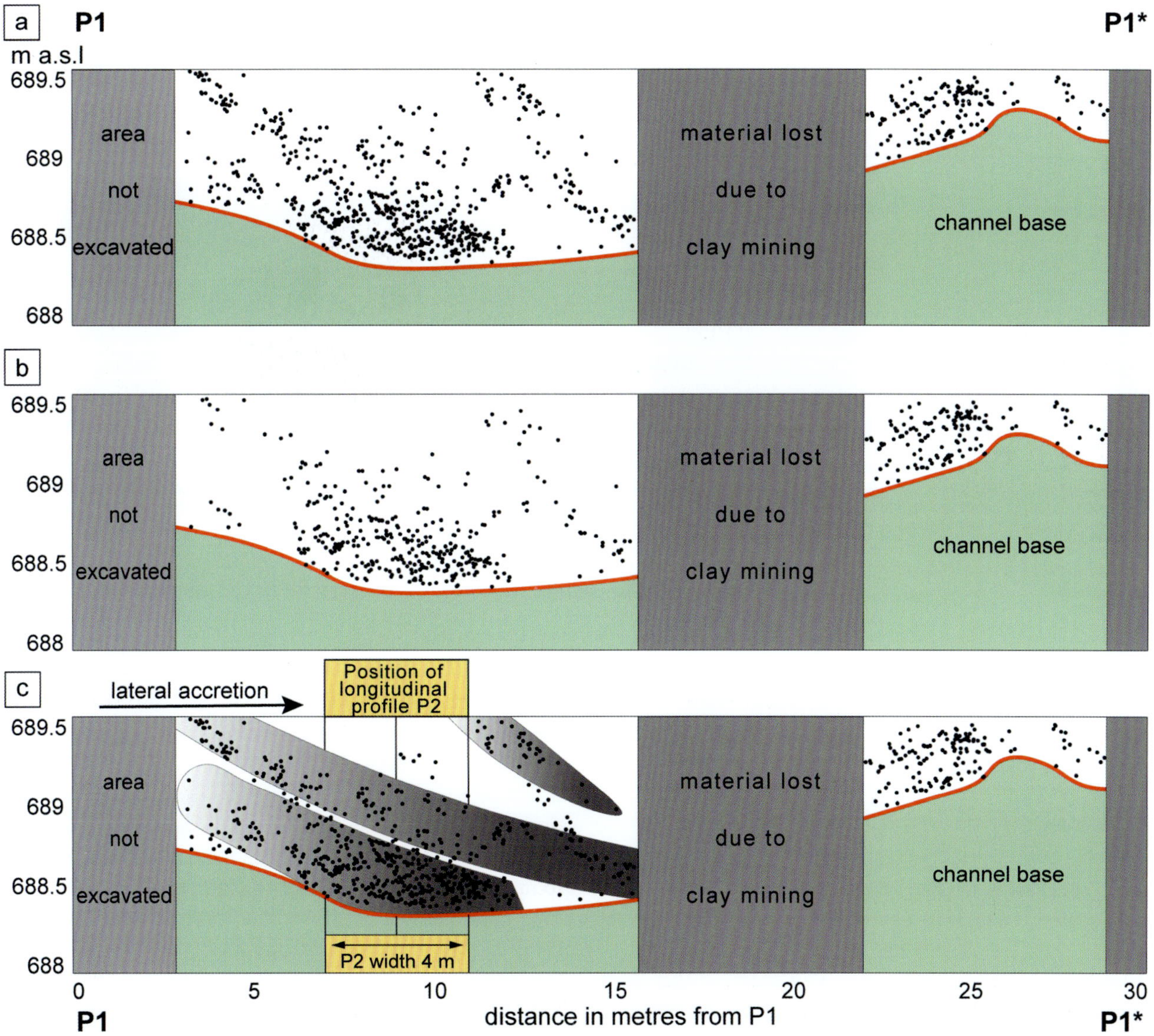

Figure 3.2.4. Accumulated transverse channel profile section in the excavation area of the local stratigraphic level HAM4 at the early Late Miocene Hammerschmiede locality. The section is west-east oriented (see Fig. 3.2.3), looking to the north and comprises data accumulated during excavations from 2017–2018 and 2019–2023 from a four metres wide transect along the median profile course from P1 (E4394793.52, N5310824.13) to P1* (E4394822.56, N5310821.53). Excavated specimens (black dots) are plotted according to their elevation above the HAM4 channel base (red line). a–b shows different size filtering of the fossil found: (a and c) fossils larger than 2 cm; (b) fossils larger than 3 cm. Note that no distinction is made between the object sizes in the right channel area. The grey area has been lost due to clay mining or has not been excavated. Grey shaded overlays indicate the location of inclined packages (dip angle range from 4° in the west to 10° in the east), interpreted as lateral accretion of a meander point bar. Easting coordinates correspond to Gauss-Krüger Zone 4 grid in metres and elevation in m a.s.l.

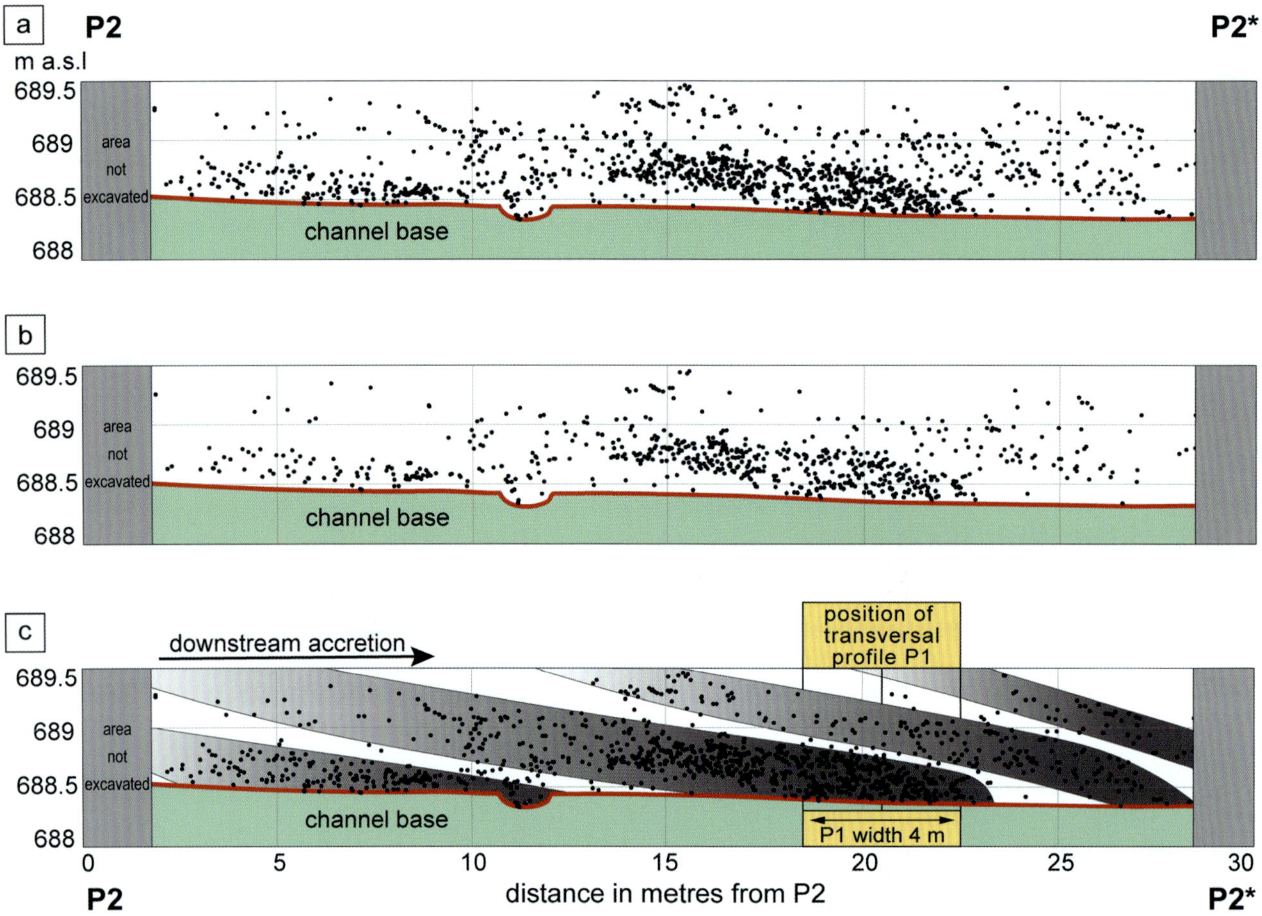

Figure 3.2.5. Accumulated longitudinal channel profile section in the excavation area of the local stratigraphic level HAM4 at the early Late Miocene Hammerschmiede locality. The section is approximately south-north oriented (see Fig. 3.2.3), looking to the north and comprises data accumulated during excavations from 2019 to 2023 from a four metres wide transect along a median profile course from P2 (E4394793.83, N5310804.81) to P2* (E4394806.88, N5310832.25). Excavated specimens (black dots) are plotted according to their elevation above the HAM4 channel base (red line). a–b shows different size filtering of the fossils found: (a and c) fossils larger than 2 cm; (b) fossils larger than 3 cm. The grey area at P2* has been lost due to clay mining or has not been excavated. Grey shaded overlays indicate the location of inclined packages (dip angle range from 2° in the south to 3° in the north), interpreted as downstream accretion of a meander point bar. Easting coordinates correspond to Gauss-Krüger Zone 4 grid in metres and elevation in m a.s.l.

material, indicating their role as fossil traps (see the undulating channel base in Fig. 5.1.1.3 and Case study 5.8.2 with Fig. 5.8.2.1).

3.4 Update of the age model

Bio-magnetostratigraphic investigations at the Hammerschmiede outcrop and a nearby research borehole revealed an age of the clay-pit succession between 11.667 and 11.42 Ma, encompassing the Middle to Late Miocene transition at 11.65 Ma (Kirscher et al. 2016). The channel HAM5 has been dated to 11.62 Ma, whereas the channel HAM4 dates to between 11.59 and 11.44 Ma. However, based on the outcrop conditions until 2015, the age model assumed a hiatus of 150 kyr, represented by the erosional structure HAM4 (see Kirscher et al. 2016, Fig. 1c and Fig. 3.0.2). The westward expansion of the mine now corroborates a spatially restricted channel structure (Fig. 3.0.3), and no erosional features west of it. Instead, the sedimentation between 688 and 693 m a. s. l. appears continuous without any hiatus. Based on this observation, we need to update the chronology of the upper part of the Hammerschmiede section. Based on the mean sedimentation rate (27 cm/kyr, Kirscher et al. 2016) and the position of the channel 10 m above HAM5 (Fig. 3.0.3), the HAM4 river is only 37 kyrs younger than the HAM5 rivulet, implying a revised age of 11.58 Ma. Accordingly, the fossiliferous horizons HAM1 and HAM7 can now be dated to 11.59 Ma, HAM2 and HAM6 to 11.57 Ma, and HAM3 to 11.56 Ma. This revision implies that the seven fossiliferous levels of Hammerschmiede encompass only 60 kyrs, from 11.62 Ma (HAM5) to 11.56 Ma (HAM3), at the very base of the Late Miocene (Tortonian, Pannonian).

Figure 3.2.6. Profile section at the base of the HAM4 channel (May 2018) at the early Late Miocene Hammerschmiede locality. The section is west-east oriented, looking to the north and roughly cuts transverse to the course of the HAM4 channel. The green overlay at the bottom of the profile marks the channel base and the transition from trough cross-bedded fine sandy sediments to the erosive base of the channel. The troughs contain coarse-grained layers with a high proportion of reworked pedogenic carbonate concretions (white spots), clay pebbles, and dark (brown) coloured bones (white arrows). Reddish or brownish layers within the trough represent heavy mineral (especially garnet) concentrations. Channel lag accumulations (unionid shells and dark spots from bones) are seen in the bottom left. Scale bar represents 10 cm.

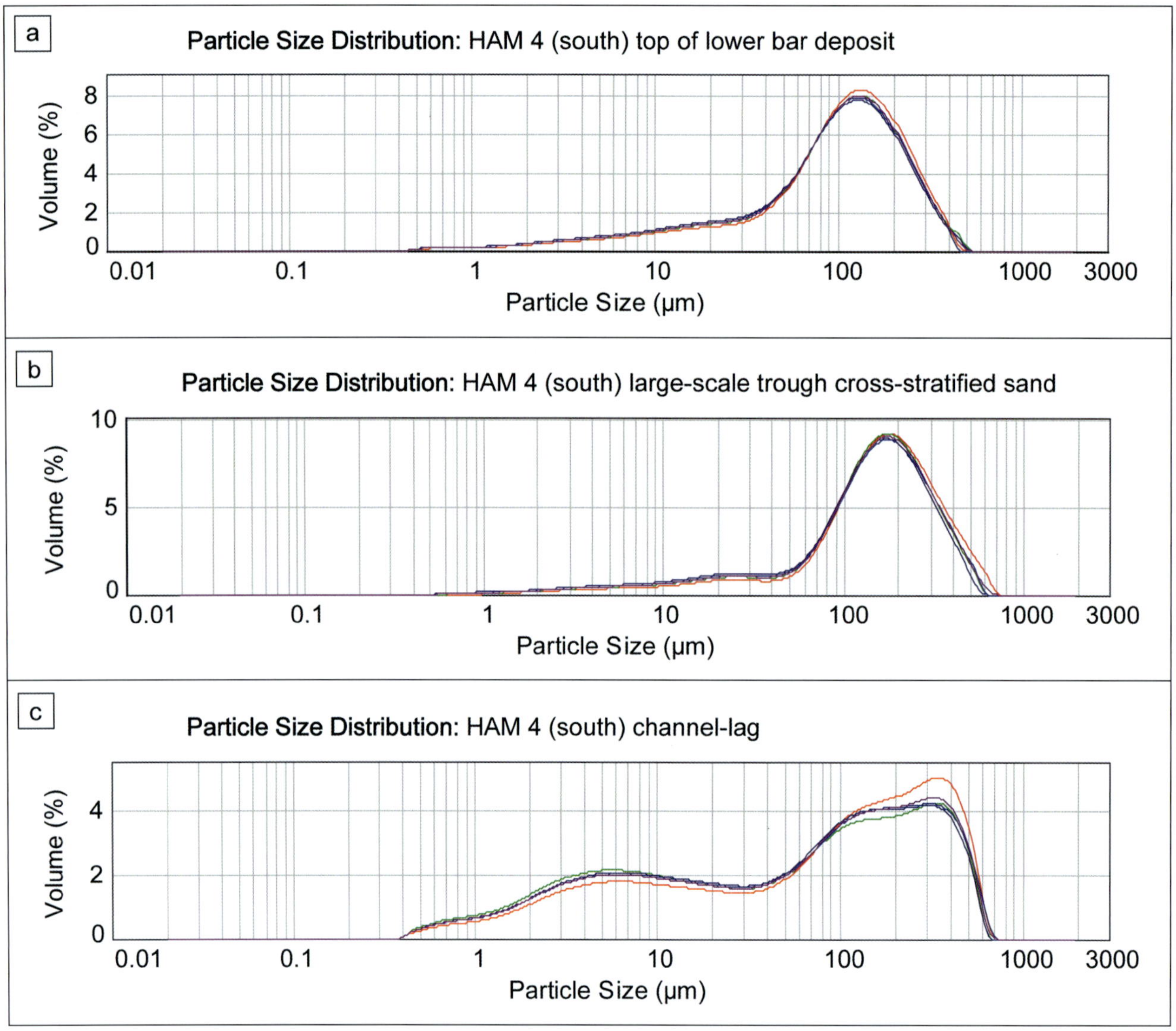

Figure 3.2.7. Particle size distributions within the excavated substrata of the HAM4 channel. Samples were taken from the 2023 excavation area of the local stratigraphic level HAM4 South at the early Late Miocene Hammerschmiede locality. (a) top of lower bar deposit, (b) large-scale trough cross-stratified sand, and (c) channel lag deposit. All samples have been decalcified before measurement using Mastersizer2000 (Malvern Panalytical). See text for discussion.

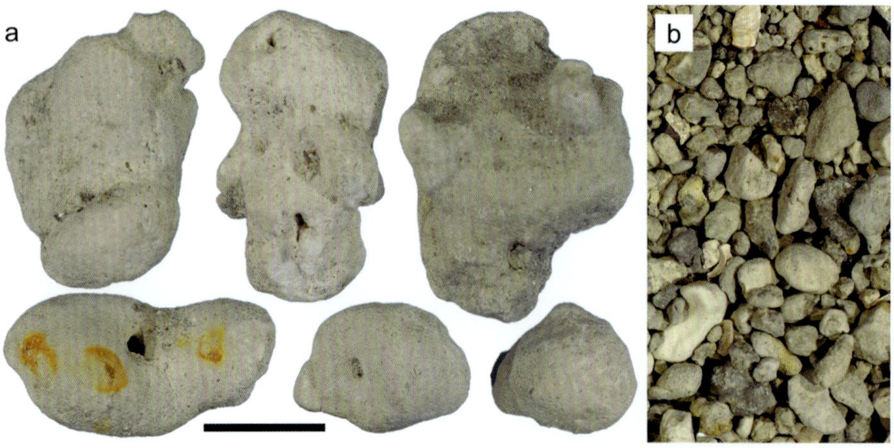

Figure 3.2.8. Reworked pedogenic carbonate concretions from the early Late Miocene Hammerschmiede site and the local stratigraphic levels HAM5 and HAM4. Particularly large (a) pedogenic carbonate nodules were found less frequently, whereas small (b) nodules occurred very frequently. Scale bar equals 10 mm.

Figure 3.2.9. Profile section in the local stratigraphic level HAM4 North (September 2024) at the early Late Miocene Hammerschmiede locality. The section is west-east oriented, looking to the north and roughly cuts transverse to the course of the HAM4 channel. The photograph shows large-scale trough cross-stratification from lower bar deposits superimposed by small-scale trough cross-stratification (upper left) and inclined bedded sands (upper right). Troughs contain numerous unionid shells, reworked pedogenic carbonate concretions (white dots), and bones (dark dots). Scale bar represents approximately 10 cm.

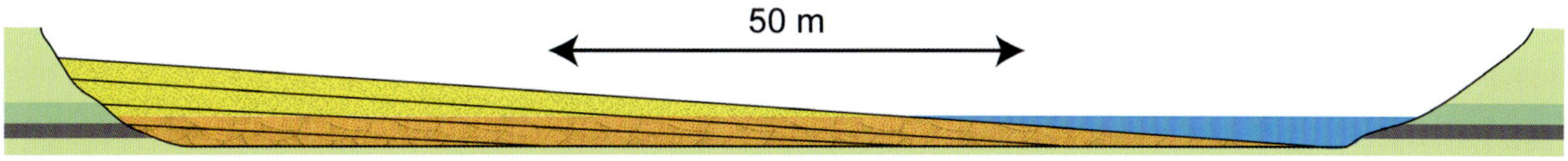

Figure 3.2.10. Idealized cross-section of the HAM4 channel during mean low-water flow shortly before river avulsion. The bedrock colours refer to Fig. 3.0.3. Lower point bar deposits are highlighted in orange and upper point bar deposits in yellow. Note that in the dry season, the point bars provide comparatively wide unvegetated sandy habitats along the stream.

4 Palaeontology – faunal list of the Hammerschmiede

A first comprehensive faunal summary was provided by Kirscher et al. (2016) which was subsequently updated by Böhme et al. (2019a). With each excavation season, the Hammerschmiede site yields additional new species, necessitating updates to the previously published lists as well as additions and corrections. Some vertebrate species are completely new to science. Others are already known from other sites. Many taxonomic groups have not yet been extensively studied, but it is clear that the fauna is characterised by remarkable richness. To date, a total of 169 taxa can be recorded in the faunal list of Hammerschmiede (HAM1–6) (see Tab. 2). An overview of the major taxonomic groups comprises 15 molluscs (3 Bivalvia and 12 Gastropoda), 4 arthropods (crayfish, mayfly, skin beetle, termite), 11 fishes, 13 amphibians, 25 reptiles (1 Choristodera, 9 Chelonia, 15 Squamata), 15 birds, 86 mammals (15 Eulipotyphla, 1 Chiroptera, 2 Primates, 28 Carnivora, 2 Proboscidea, 5 Perissodactyla, 7 Artiodactyla, 2 Lagomorpha, and 24 Rodentia).

Table 2. Faunal list of the Hammerschmiede locality and the local stratigraphic levels HAM5, HAM4, HAM6, and HAM1–3 (combined). For species marked with *, Hammerschmiede is the type locality of the holotypes. Updated and supplemented with new expertise based on Kirscher et al. (2016), Table 1 and Böhme et al. (2019a), Supplementary Table 1.

Order	Family	Taxon	HAM 5	HAM 4	HAM 6	HAM 1–3
Unionida	Margaritiferidae	*Margaritifera (Pseudunio) flabellata*	X	X		X
Sphaeriida	Sphaeriidae	*Sphaerium (Amesoda) rivicola*	X			X
		Pisidium (Pisidium) amnicum	X	X		X
Architaenioglossa	Bithyniidae	*Bithynia* sp. 1	sp.	sp.		X
		Bithynia sp. 2				X
Heterobranchia	Valvatidae	*Borysthenia* sp.				X
Hygrophila	Planorbidae	*Segmentina* sp.				X
		Gyraulus sp.	?	?		X
		Planorbarius sp.		X		
Ellobiida	Carichiidae	*Carychium* sp.				X
Stylommatophora	Strobilopsidae	*Strobilops* sp.				X
	Filholidae	*Triptychia* sp.		X		X
	Helicidae	*Cepaea* sp.	X	X		
	Elonidae	*Tropidomphalus* sp.	X	X		
	Limacidae	*Limax* sp.	X	X		X
Decapoda	Potamidae	*Potamon* sp.			X	
Coleoptera	Dermestidae	gen. et sp. indet.		X		
Ephemoptera	incertae sedis	ichnosp. *Rhizocorallium jenense*		X		
Blattodea	Archotermopsidae	cf. *Microcarpolithes hexagonalis*	X	X		
Esociformes	Esocidae	*Esox* cf. *lepidotus*	X	X		X
Siluriformes	Siluridae	*Silurus* cf. *joergi*	X	X	X	X
Cypriniformes	Cyprinidae	*Tinca* sp.	X	X		X
		Gobio sp.		X		
		Leuciscus sp.	X	X		X
		Luciobarbus sp.	X	X		
	Cobitidae	*Cobitis* sp.	X	X		X
Perciformes	Gobiidae	*Gobius* sp.	X	X		X
	Percidae	*Perca* sp. 1	X	X		indet.
		Perca sp. 2	X	X		
Salmoniformes	Salmonidae	*Hucho hucho*	X	X		
Urodela	Proteidae	*Mioproteus* sp.	X	X	X	X
	Cryptobranchidae	*Andrias scheuchzeri*	X	X		X
	Scapherpetontidae	gen. et sp. indet.	X	X		X
	Batrachosauroididae	gen. et sp. indet.	X	X		X
	Salamandridae	*Chelotriton* sp.	X	X	X	X
		Triturus roehrsi	sp.	sp.		X
		Triturus aff. *montandoni*				X

Order	Family	Taxon	HAM 5	HAM 4	HAM 6	HAM 1–3
Anura	Palaeobatrachidae	*Palaeobatrachus* sp.		X		X
	Discoglossidae	*Latonia gigantea*	X	X	X	X
	Bufonidae	*Bufo* cf. *viridis*	X	sp.	X	X
	Hylidae	*Hyla* sp.	X	X		X
	Ranidae	*Pelophylax* sp.	X	X		X
	Pelobatidae	*Eopelobates* sp.	X	X		X
Choristodera		*Lazarussuchus* sp.	X	X		X
Chelonia	Trionychidae	*Trionyx* sp.	X	X	X	X
	Chelydridae	*Chelydropsis* sp.	X	X	X	X
	Geoemydidae	*Clemmydopsis* sp.	X	X	X	X
		Mauremys sarmatica	X	X	X	
		Melanochelys sp.		X		
		Ptychogaster (T.) batalleri		X		
	Testudinidae	*Testudo* sp.	X	X		X
		Chersine sp.		X		
	incertae sedis	gen. et sp. indet.		X		
Squamata	Chamaeleonidae	*Chamaeleo* sp.		X		
	Lacertidae	*Lacerta* sp. 1	X	indet.		X
		Lacerta sp. 2		X		X
	Scincidae	*Chalcides* sp.		X		X
		Lygosominae gen. et sp. indet.		X		
		Scincomorpha gen. et sp. indet.		X		
	Anguidae	*Pseudopus pannonicus*	X	X	X	
		Ophisaurus sp.	X	X		
		Anguis sp.	X	X		
	Boidae	Erycinae gen. et sp. indet.	X			
	Amphisbaenidae	gen. et sp. indet.		X		X
	Colubridae	Colubrinae gen. et sp. indet. 1	X	X	X	X
		Colubrinae gen. et sp. indet. 2	X	X		X
		Natricinae gen. et sp. indet. 1	X	X		X
		Natricinae gen. et sp. indet. 2	X			X
Gruiformes	Gruidae	gen. et sp. indet.		X		
Anseriformes	Anatidae	*Allgoviachen tortonica**		X		
		cf. *Mioquerquedula* sp.		X		
		gen. et sp. indet. 1	X	X		
		gen. et sp. indet. 2		X		
Accipitriformes	indet.	gen. et sp. indet. 1	X	X		
		gen. et sp. indet. 2		X		
Strigiformes	Strigidae	gen. et sp. indet.	X	X		
Suliformes	Anhingidae	*Anhinga pannonica*	X	X		
	Phalacrocoracidae	cf. *Microcarbo* sp.		X		
Galliformes	Phasianidae	cf. *Miophasianus altus*		X		
	indet.	gen. et sp. indet. 1	X			
		gen. et sp. indet. 2		X		
Coraciiformes	Alcedinae	gen. et sp. indet.		X		
Passeriformes	Corvidae	cf. *Pica* sp.		X		
Eulipotyphla	Erinaceidae	*Galerix exilis*	indet.	X		X
		? *Schizogalerix* sp.		X		
	Plesiosoricidae	*Plesiosorex schaffneri*	X	X	X	X
	Talpidae	*Gerhardstorchia quinquecuspidata*	X	div. sp.	indet.	X
		Desmanella sp.	X			
		Talpa sp.	X			
		Proscapanus sp.	X			
	Dimylidae	*Plesiodimylus johanni*	X	X		
		Metacordylodon schlosseri	X	X		

Table 2 continued on next page ▶

◄ *Table 2 continued*

Order	Family	Taxon	HAM 5	HAM 4	HAM 6	HAM 1–3
Eulipotyphla (continued)	Soricidae	*Crusafontina exculta**	X	X		X
		Paenelimnoecus crouzeli	X			
		Dinosorex n. sp.	X	X	X	
		Paenesorex sp.				X
		gen. et sp. indet. 1	X	X		
		gen. et sp. indet. 2	X			
Chiroptera		Chiroptera div. sp.	X	X		
Primates	Hominidae	*Danuvius guggenmosi**	X	X		
		*Buronius manfredschmidi**	X			
Carnivora	Amphicyonidae	gen. et sp. indet.	X	X	X	
	Ursidae	*Kretzoiarctos beatrix*	X	X		
	Ailuridae	*Alopecocyon goeriachensis*	X	X		
		Simocyoninae gen. et sp. indet.		X		
	Mustelidae/Guloninae	*Martes sansaniensis*	X	X		
		Martes munki	X	X		
		Martes sp.				X
		*Circamustela hartmanni**	X	X		X
		Laphyctis mustelinus		X		
		Guloninae gen. et sp. indet.	X	X		
	Mustelidae/Lutrinae	*Paralutra jaegeri*	X	X		
		*Vishnuonyx neptuni**		X		
		Lartetictis cf. *dubia*		X		
	Mustelidae/Leptarctinae	*Trocharion albanense*	X	X		
	Mustelidae/Mellivorinae	*Eomellivora moralesi*	X			
	Mephitidae	*Proputorius sansaniensis*				X
		Proputorius pusillus				X
		Palaeomeles pachecoi	X	X		
	Hyaenidae	*Thalassictis montadai*	X	X	X	
		gen. et sp. indet.	X			
	Viverridae	*Semigenetta sansaniensis*	X	X		X
		Semigenetta grandis		X		
		Viverrictis modica		X		
	Felidae	*Pseudaelurus quadridentatus*	X			
		Metailurini indet.	X			
	Barbourofelidae	*Sansanosmilus* sp.	X	X		
	Semantoridae	*Potamotherium* sp.	X	X		
	Phocidae	gen. et sp. indet.	X	X		
Proboscidea	Gomphotheriidae	*Tetralophodon longirostris*	X		X	
	Deinotheriidae	*Deinotherium levius*	X	X		
Perissodactyla	Rhinocerotidae	*Hoploaceratherium belvederense*	X	X		
		Aceratherium sp.	X	X		
	Chalicotheriidae	Schizotheriinae gen. et sp. indet.		X		
		Chalicotheriinae, *Anisodon* sp.	X			
	Equidae	*Sinohippus* sp.	X	X		
Artiodactyla	Suidae	*Listriodon splendens*	X	X	X	
		Parachleuastochoerus steinheimensis	X	X	X	X
		Propotamochoerus palaeochoerus		X		
	Tragulidae	*Dorcatherium naui*	X	X		
	Moschidae	gen. et sp. indet.	X	X		
	Cervidae	*Euprox furcatus*	X	X	indet.	
	Bovidae	*Miotragocerus monacensis*	X	X	X	
Lagomorpha	Ochotonidae	*Prolagus oeningensis*	X	X		
		Eurolagus fontannesi	X	X		X
Rodentia	Sciuridae	*Csakvaromys bredai*	X	X		X
		Albanensia grimmi	X	X		X
		Blackia miocaenica	X	X		

Order	Family	Taxon	HAM 5	HAM 4	HAM 6	HAM 1–3
Rodentia (continued)	Sciuridae	*Miopetaurista* sp.	X	?		
	Castoridae	*Euroxenomys minutus*	X	X		X
		Steneofiber depereti	X	X		X
	Gliridae	*Microdyromys complicatus*	X			
		Muscardinus hispanicus	X	X		X
		Muscardinus sp.	X	X		
		Bransatoglis sp.	X			
		Glirulus conjunctus	X	?		X
		Eliomys reductus	sp.			X
		Eliomys assimilis				X
		Myoglis meini	X	X		X
	Eomyidae	*Eomyops catalaunicus*	X			X
		Keramidomys sp.	X			
	Cricetidae	*Democricetodon* n. sp.	X	X		X
		Collimys hiri	X	X		X
		Megacricetodon minutus	X	X		X
		*Microtocricetus molassicus**	X	X		X
		Eumyarion latior	X	X		
		Cricetodontini gen. et sp. indet.		X		
	Platacanthomyidae	*Neocometes* sp.		X		
	Anomalomyidae	*Anomalomys gaudryi*	X	X	X	X

5 Taphonomy

5.1 Area of excavation and documentation

The excavations were always clearly structured and documented separately according to their respective stratigraphic position in the profile. The local stratigraphic levels HAM5 and HAM4 should be regarded as separate sites representing different time periods and are treated here sequentially.

5.1.1 HAM5

In the years 2011 to 2015, no extensive excavations could take place. The exposure conditions of HAM5 were located on the steep slope of the active clay mining level, allowing only smaller-scale excavations to be carried out. In the excavation year 2016, a larger HAM5 area was exposed for the first time due to mining, as the overlying sediment was now removed. Here, for the first time, it was possible to excavate over three weeks and to work on an area of approximately 10 m^2. The finds from this period can only be estimated within the excavation area by the chronological sequence of finds. Even before the 2016 summer excavation season, a significant portion of HAM5 would have been lost to clay mining, as mining of the deeper layers was now to take place. Here, approximately 23 tonnes of the find layer was recovered on May 17, 2016, with an excavator and stored for later processing. In the subsequent excavation, HAM5 was explored in a further section with a corresponding gap in areas from previous years and also only rough localisation of finds. Only from 2017 onwards was it possible to begin a coherent excavation of HAM5, which also links up with a gap to previous areas (Fig. 5.1.1.1). Starting in 2017, finds and features were spatially surveyed, and exact find positions can now be placed in relation to each other. This was followed by excavations in 2017 (40 m^2), 2018 (100 m^2), 2020 (50 m^2), 2021 (100 m^2), 2022 (2 m^2), and 2024 (37 m^2). No excavations took place at HAM5 in 2019 and 2023.

Currently, the overlying strata of HAM4 North are blocking further progress, so that the documentation of the overlying HAM4 is necessary as preliminary work in order to continue HAM5 excavations. For the excavated areas and years 2017–2018, 2020–2022, and 2024 see Fig. 5.1.1.2, and for corresponding channel base morphology at the sections see Fig. 5.1.1.3.

5.1.2 HAM4

The local stratigraphic level HAM4 was already visible in the slope since the beginning of current research efforts in 2008. Due to its location in the middle of the mining slope, there was no possibility to realise larger

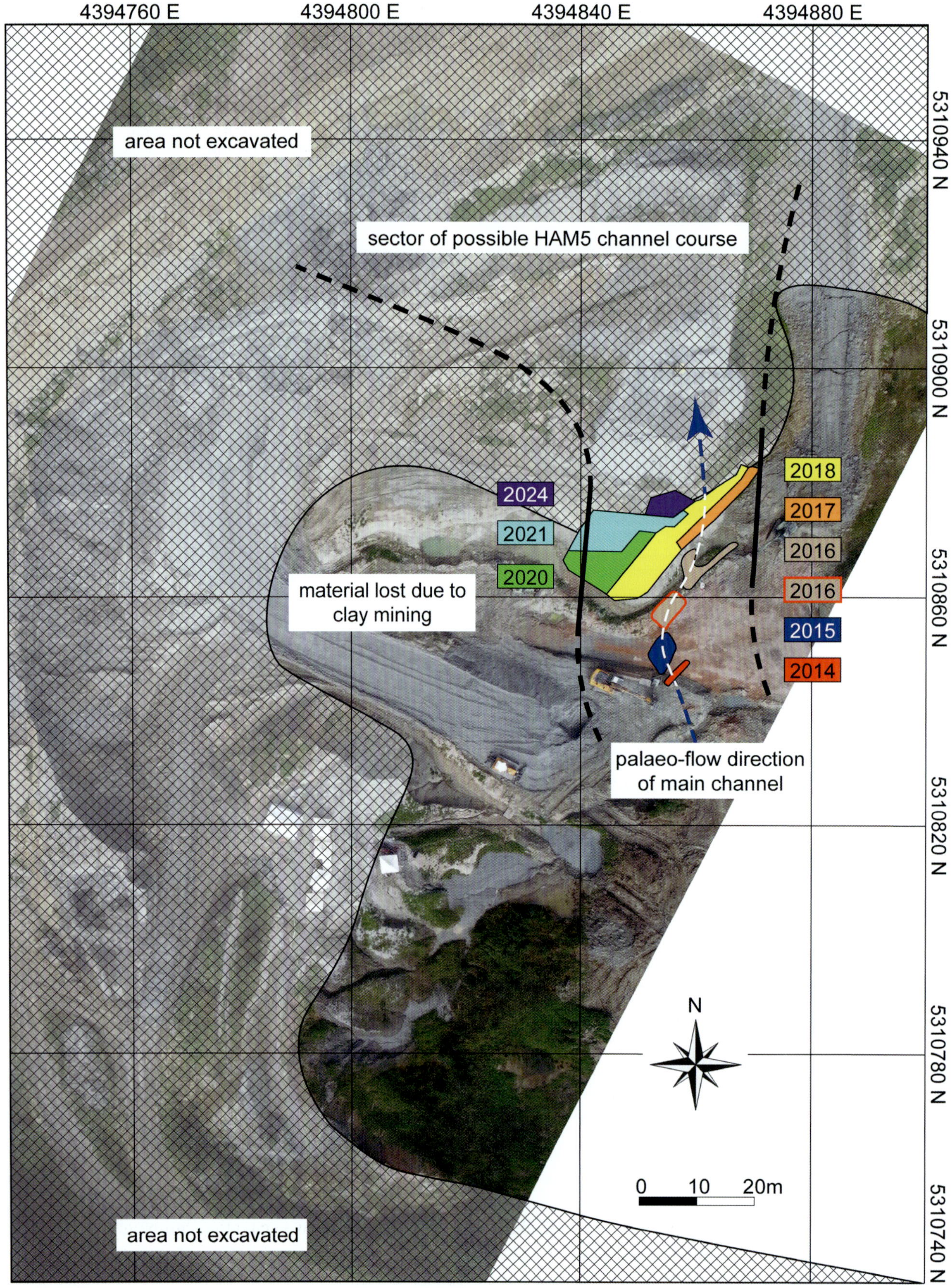
4394760 E
4394800 E
4394840 E
4394880 E
5310940 N
5310900 N
5310860 N
5310820 N
5310780 N
5310740 N
area not excavated
sector of possible HAM5 channel course
2024
2021
2020
2018
2017
2016
2016
2015
2014
material lost due to clay mining
palaeo-flow direction of main channel
N
0 10 20m
area not excavated

◄ **Figure 5.1.1.1.** Top view of the Hammerschmiede clay pit (drone image, September 2022) including the excavation areas within the local stratigraphic level HAM5 at the early Late Miocene Hammerschmiede locality. The areas of 2017, 2018, 2020, 2021, and 2024 mark contiguously excavated and spatially documented areas. The areas of 2014, 2015, and 2016 correspond to approximate estimates. The red encircled 2016 polygon indicates an area that was under threat of destruction from mining activity. Sediments of this section have been removed, stored (approximately 23 tonnes), and separately processed over the following years. Thick black lines indicate the channel course based on outcrop observations. Dashed thick lines indicate the presumed course from the south and a possible area of the course to the north (secured by laterally delimiting outcrops). The thick blue and white dashed lines indicate the course of the deepest incised main channel of HAM5. Unexcavated areas are shaded and other areas at this elevation have been lost due to clay mining. Coordinates correspond to Gauss-Krüger Zone 4 grid in metres. Grid spacing equals 40 metres.

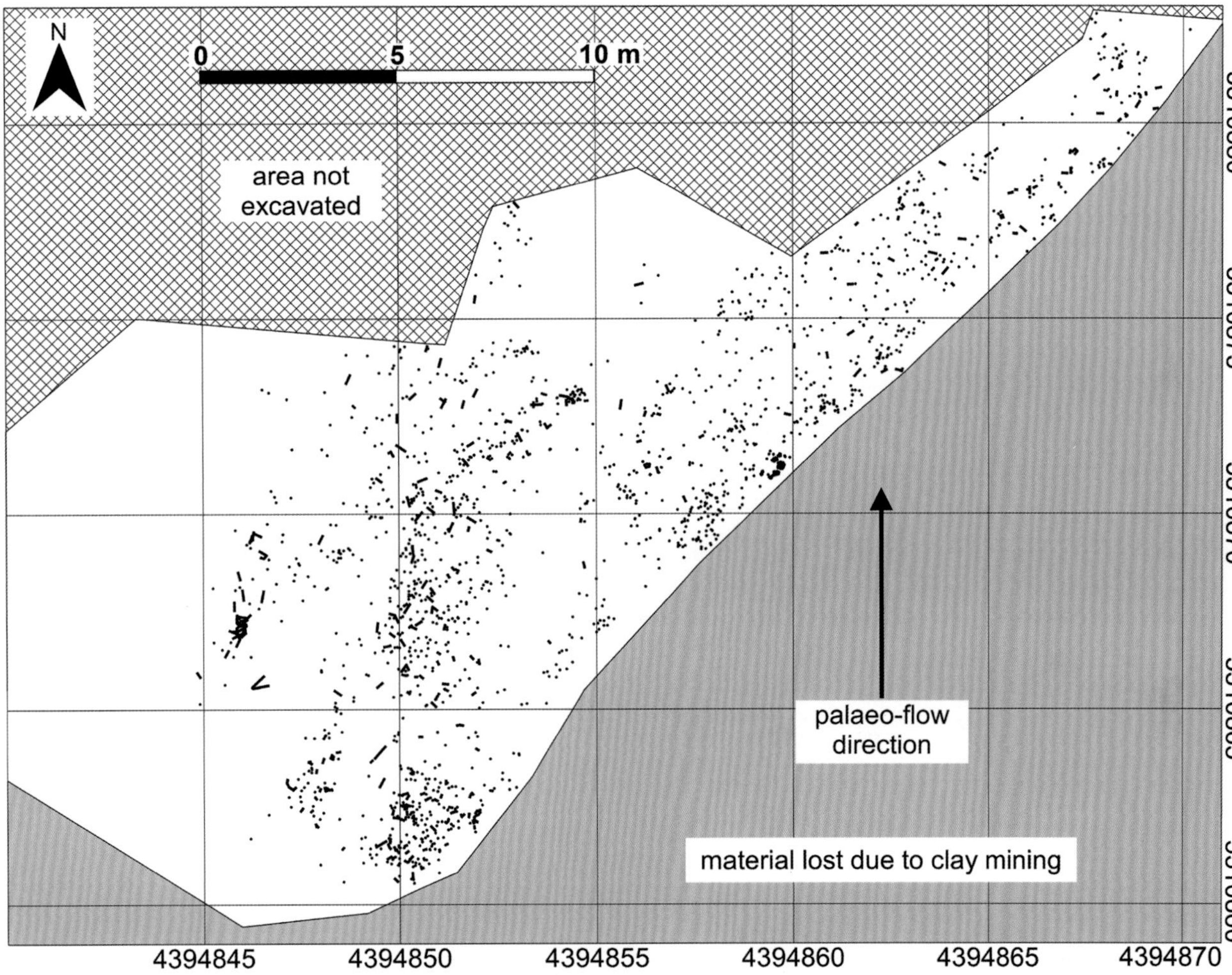

Figure 5.1.1.2. Detailed map of find distributions in the area of the local stratigraphic level HAM5 at the early Late Miocene Hammerschmiede site. The white polygon frames fully excavated areas of the years 2017, 2018, 2020, 2021, 2022, and 2024 (no excavations took place in 2019 and 2023). Black dots mark excavated and measured isolated finds (mostly bones). Black lines represent elongated bones including length and orientation. The black arrow indicates the presumed palaeo-flow direction of the watercourse. The grey area has been lost due to clay mining, and the shaded area includes unexcavated areas. Coordinates correspond to Gauss-Krüger Zone 4 grid in metres. Grid spacing equals 5 metres.

excavations until 2017. Until then, only single finds could be recovered from the slope. Only a change in the mining strategy of the pit operator (the pit became too deep, as the sediment thickness increased in the mining direction) with the creation of a mining floor in the stratigraphic level of HAM4, made it possible there to excavate over a wide area (Fig. 5.1.2.1).

After the intended end of the excavations in 2017, the extreme density of finds in HAM4, which was about to be mined, was recognised. The first finds were so promising that a new excavation campaign was carried out for a further 3 weeks in the autumn of the same year and continued in the same area in the spring of 2018. Until the following excavation period, the clay mining was so extensive that our work could only continue with a sufficiently large gap to the adjacent areas in the south, middle, and north. Until the temporary protection granted from spring 2020 and the final pro-

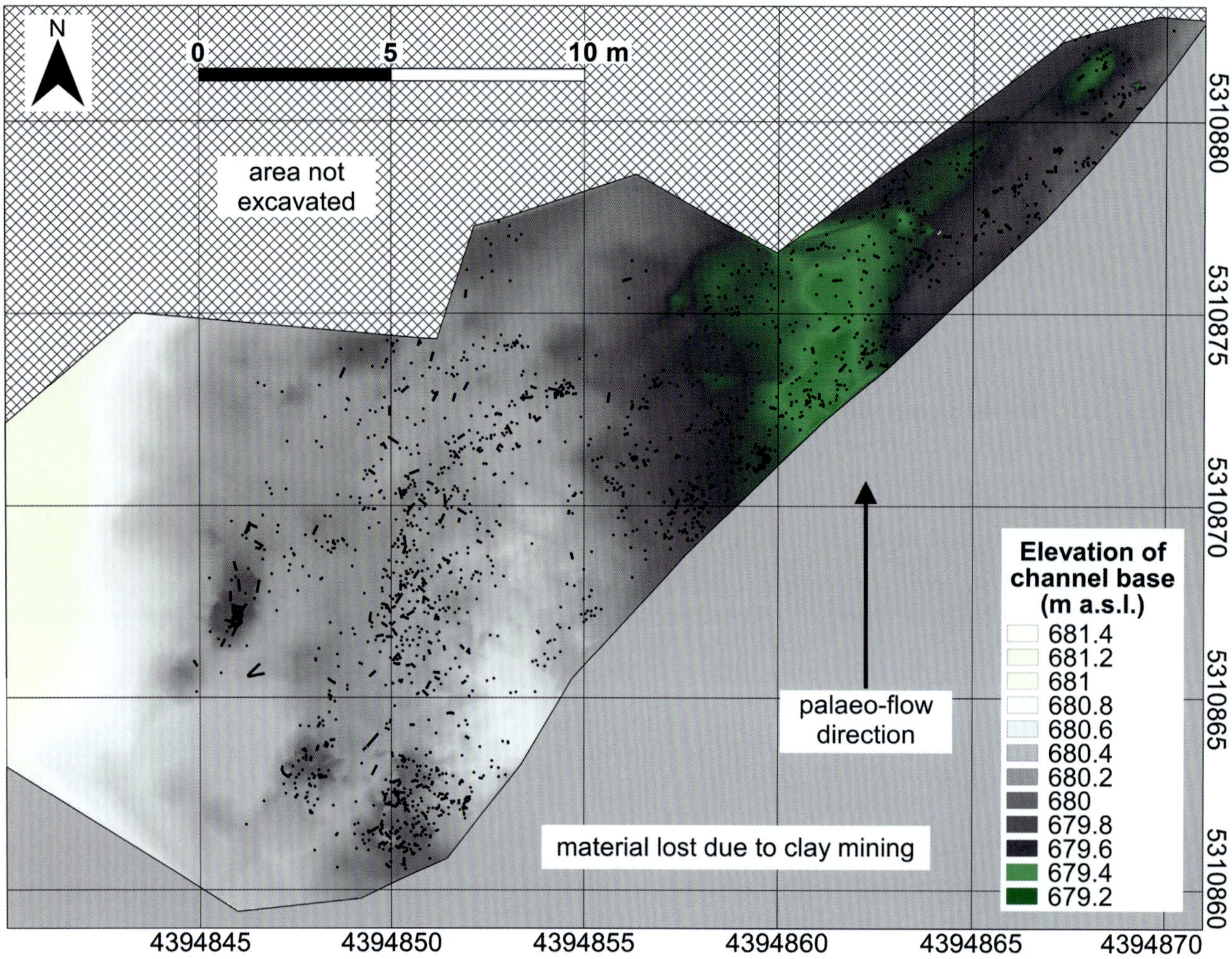

Figure 5.1.1.3. Relief map of the fossil HAM5 channel base elevation at the early Late Miocene Hammerschmiede site. The polygon frames the excavation areas of the years 2017, 2018, 2020, 2021, 2022, and 2024. Black dots mark excavated and measured isolated finds (mostly bones). The map shows the maximum erosion depth of the HAM5 channel based on measured points selected depending on the basement surface texture. The reconstructed palaeo-flow direction (black arrow) is based on the course of the deepest channel base regions (darker colours) and the dip of the channel margin (brighter colours), supplemented by the orientation of elongated objects found (black lines). The grey area has been lost due to clay mining, and the shaded area includes unexcavated areas. Values for altitude in metres above sea level. Coordinates correspond to Gauss-Krüger Zone 4 grid in metres. Grid spacing equals 5 metres.

tection declaration as a natural monument in 2021 by the district administration of Ostallgäu, large areas of the eastern rim of HAM4 and the central pit area were lost due to clay mining. The 2017 and 2018 sites are thus contiguous sites separated by respective gaps from later excavations (2017: 56 m^2 and 2018: 70 m^2; Figs 5.1.2.2 and 5.1.2.3).

This was followed by an excavation at the southern edge of the pit 2019/1 (143 m^2) (Figs 5.1.2.4 and 5.1.2.5). In the same year, in a second campaign, an excavation was carried out in the central pit area not far from the areas of the two previous years. Finally, it was possible to continue the previous year's excavations without any losses and to excavate the largest contiguous area of HAM4 to date (2019/2: 54 m^2, 2020: 60 m^2, 2021: 70 m^2, 2022-1: 140 m^2; Figs 5.1.2.2 and 5.1.2.3). In the 2022 field season, excavation continued for the first time north of the large central gap of HAM4, that was lost due to clay mining to regain access to HAM5 below (2022-2: 50 m^2, 2023: 85 m^2, 2024: 132 m^2; Figs 5.1.2.6 and 5.1.2.7).

5.2 Preserved material types

The preservation conditions in both layers of HAM5 and HAM4 are particularly favorable for the conservation of carbonate and phosphatic fossils. Bones and teeth of vertebrates especially show excellent preservation, whereas molluscs show different types of preservation. Freshwater pearl mussels (*Margaritifera* (*Pseudunio*) *flabellata*), for example, are preserved in HAM5 as internal casts possibly preserving the periostracum without the original calcareous shell,

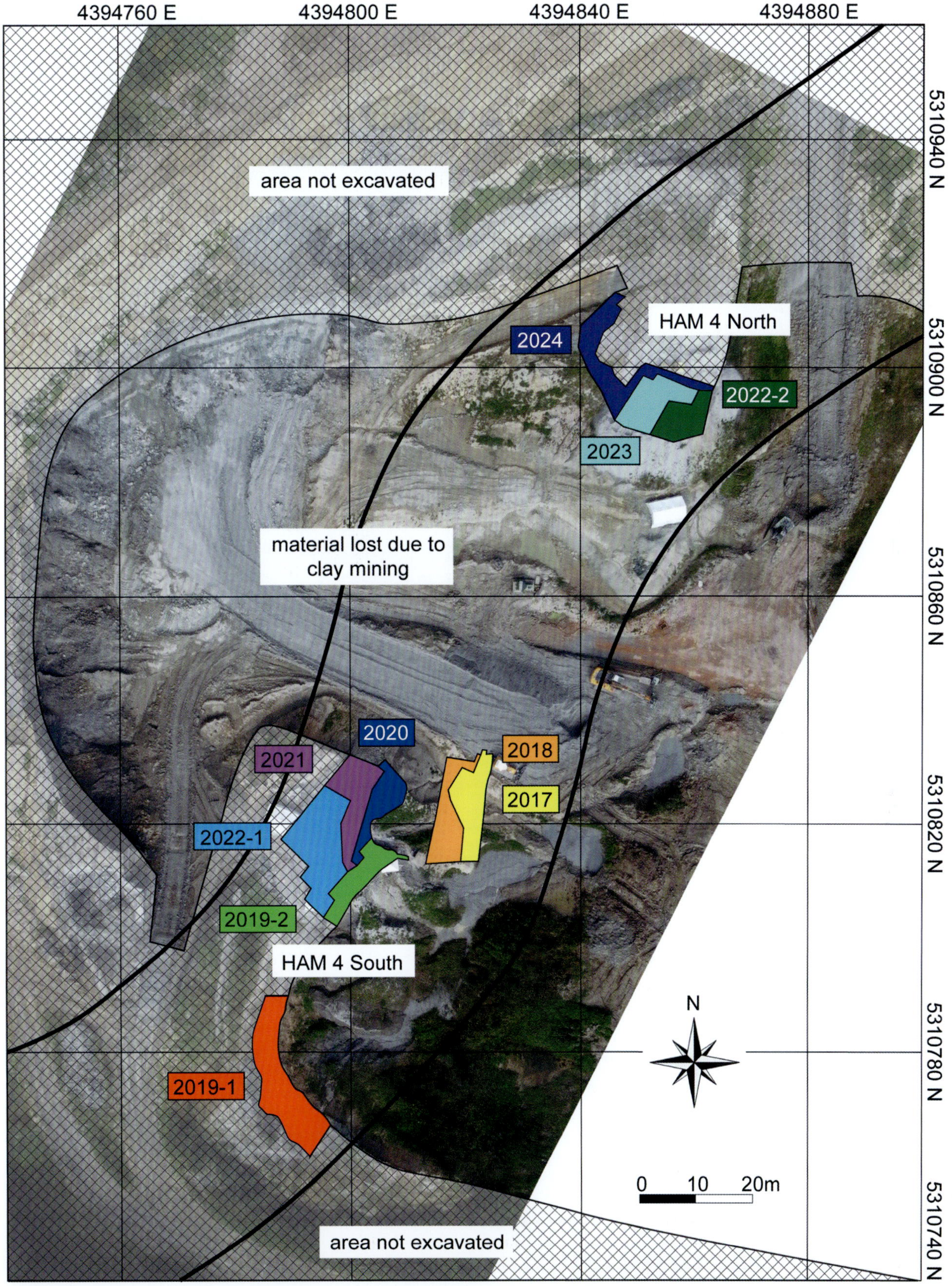

Figure 5.1.2.1. Top view of the Hammerschmiede clay pit (drone image, September 2022) including the excavation areas within the local stratigraphic level HAM4 at the early Late Miocene Hammerschmiede locality. The area between the excavation sector HAM4 North and HAM4 South and to the east of those has been lost due to mining activities. The excavated and spatially documented areas of the years 2017 to 2024 are marked by coloured areas. Thick black lines indicate the potential slightly s-shaped meandering channel course based on field observations at the mining slopes. Shaded areas mark unexcavated areas. Coordinates correspond to Gauss-Krüger Zone 4 grid in metres. Grid spacing equals 40 metres.

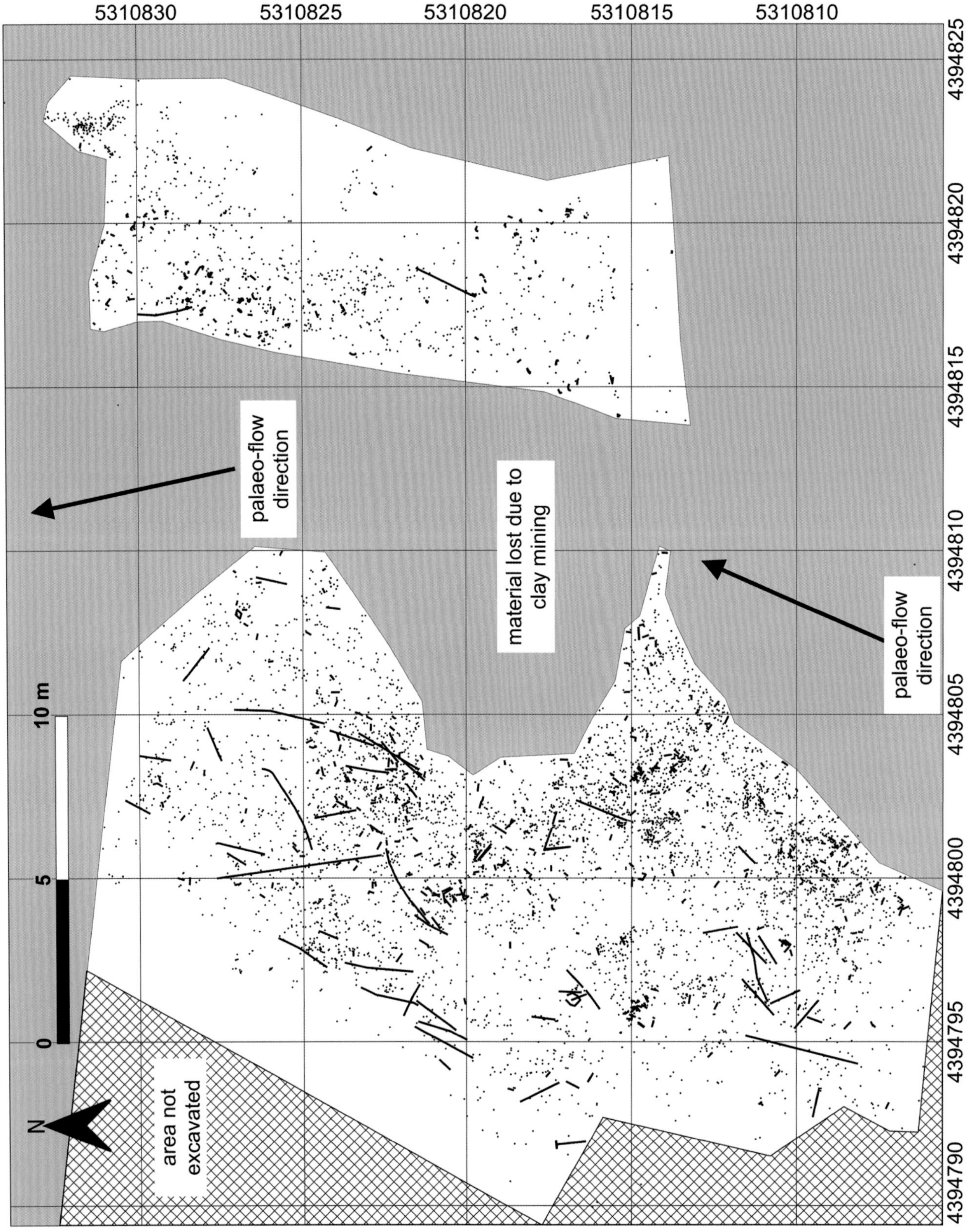

Figure 5.1.2.2. Detailed map of find distributions in the central excavation areas of the local stratigraphic level HAM4 at the early Late Miocene Hammerschmiede site. The eastern polygon (right) frames the contiguous excavation area from 2017 to 2018. The western frame (left) includes the excavation areas 2019-2, 2020, 2021, and 2022-1. Black dots mark excavated and measured isolated finds (mostly bones). Black lines represent elongated objects (the longest represent wood finds) including length and orientation. Black arrows indicate the presumed palaeo-flow direction of the HAM4 river and refer primarily to the orientation of the elongated objects found. The grey area has been lost due to clay mining, and the shaded area includes unexcavated areas. Coordinates correspond to Gauss-Krüger Zone 4 grid in metres. Grid spacing equals 5 metres.

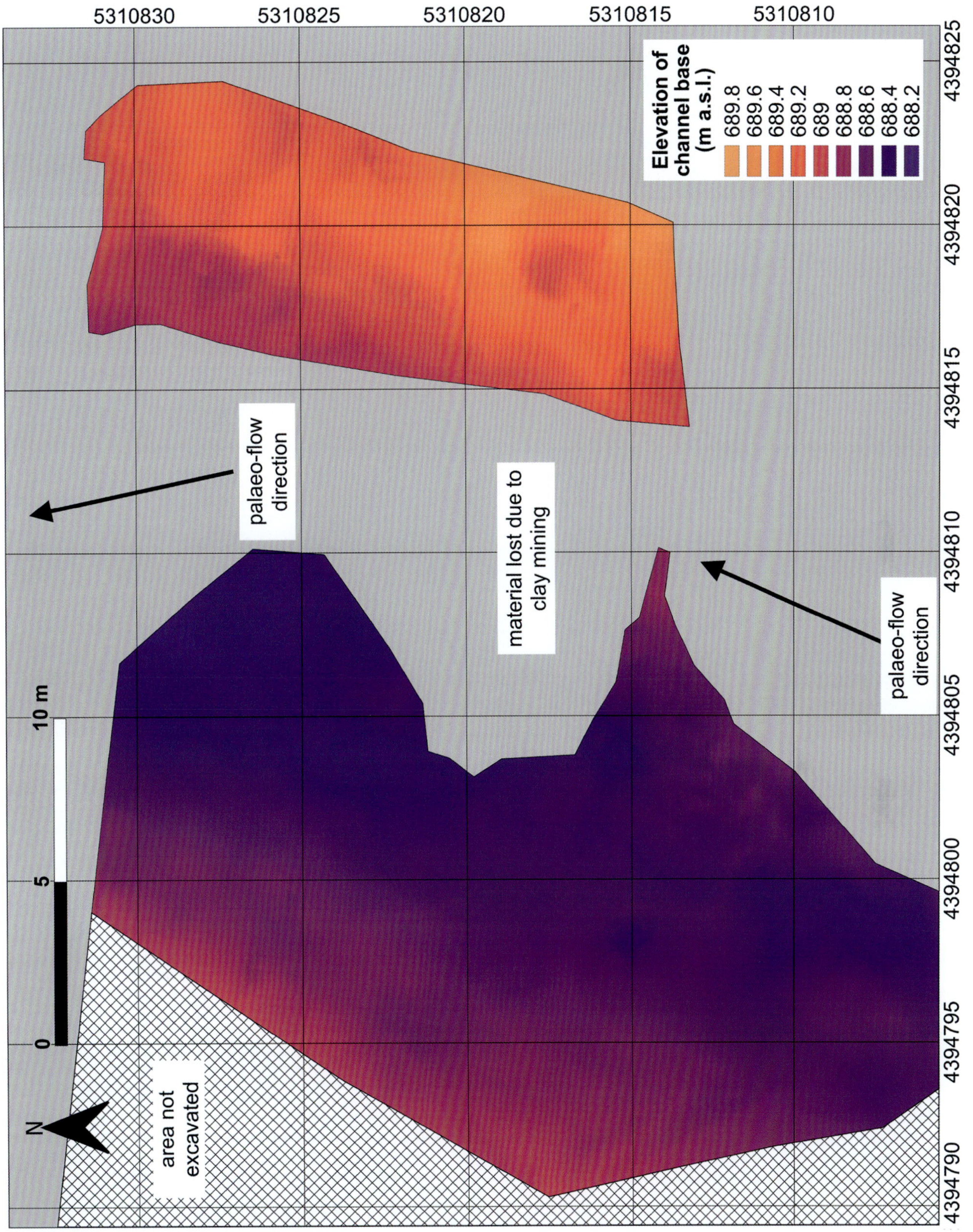

Figure 5.1.2.3. Relief map of the fossil river channel base elevation in the central excavation area of the local stratigraphic level HAM4 at the early Late Miocene Hammerschmiede site. The eastern polygon (right) frames the contiguous excavation area from 2017 to 2018. The western frame (left) includes the excavation areas 2019-2, 2020, 2021, and 2022-1. The map shows the maximum erosion depth of the HAM4 channel based on measured points selected depending on the basement surface texture. The reconstructed palaeo-flow direction (arrows) is based on the course of the deepest channel base regions (darker colours) and the dip of the channel margin (brighter colours), supplemented by the orientation of elongated objects found. The greyed area has been lost due to clay mining, and the shaded area includes unexcavated areas. Values for the altitude in metres above sea level. Coordinates correspond to Gauss-Krüger Zone 4 grid in metres. Grid spacing equals 5 metres.

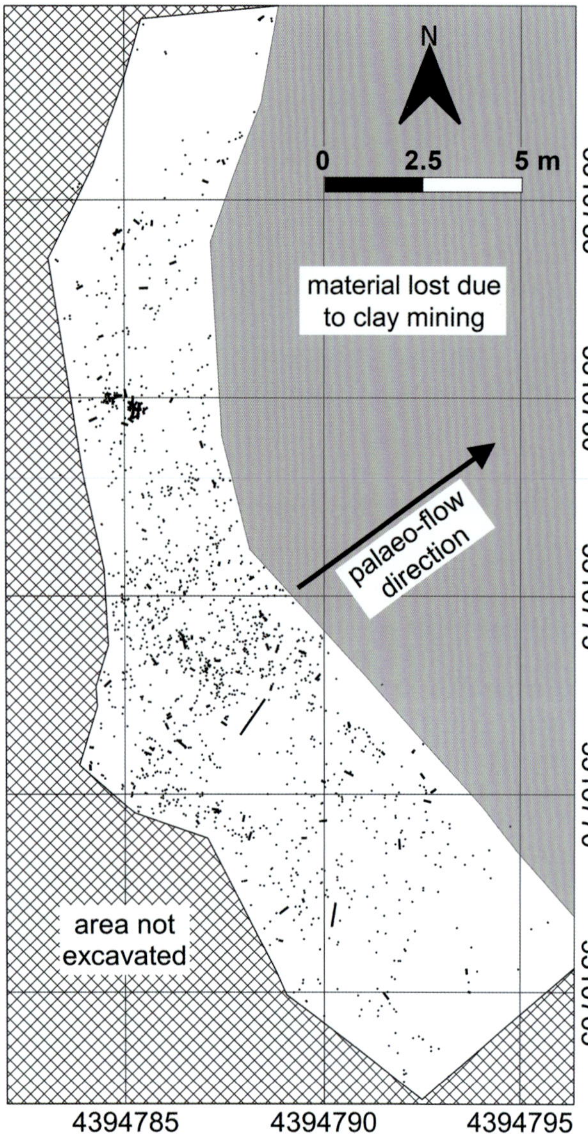

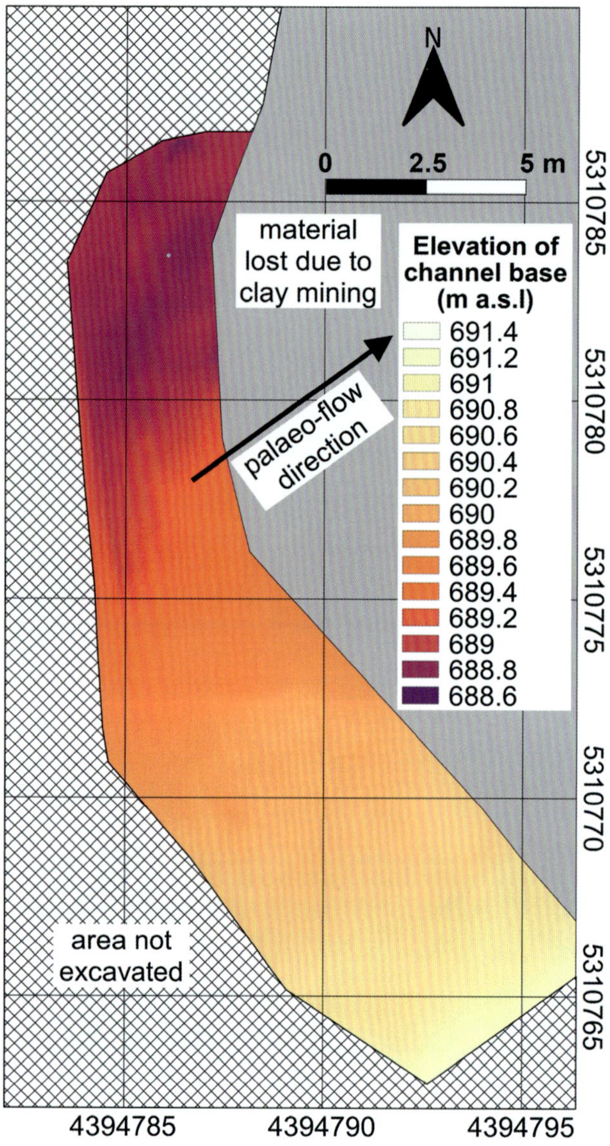

Figure 5.1.2.4. (left) Detailed map of find distribution in the southernmost excavation area of the local stratigraphic level HAM4 (2019-1) at the early Late Miocene Hammerschmiede site. Black dots mark excavated and measured isolated finds (mostly bones). Black lines represent elongated objects (the longest represent wood finds) including length and orientation. The black arrow indicates the presumed palaeo-flow direction of the HAM4 river and refers primarily to the orientation of the elongated objects found. The grey area has been lost due to clay mining, and the shaded area includes unexcavated areas. Coordinates correspond to Gauss-Krüger Zone 4 grid in metres. Grid spacing equals 5 metres.

Figure 5.1.2.5. (right) Relief map of the fossil river channel base elevation in the southernmost excavation area of the local stratigraphic level HAM4 (2019-1) at the early Late Miocene Hammerschmiede site. The map shows the maximum erosion depth of the HAM4 channel based on measured points selected depending on the basement surface texture. The reconstructed palaeo-flow direction (arrow) is based on the course of the deepest channel base regions (darker colours) and the dip of the channel margin (brighter colours), supplemented by the orientation of elongated objects found. The greyed area has been lost due to clay mining, and the shaded area includes unexcavated areas. Values for the altitude in metres above sea level. Coordinates correspond to Gauss-Krüger Zone 4 grid in metres. Grid spacing equals 5 metres.

whereas in HAM4 bivalve and gastropod shells are preserved, and some shells display a mother-of-pearl lustre. In the HAM5 clays, in some gastropod specimens the shell is almost missing, whereas in some cases in the wet state the original shell colour pattern is still recognisable (see Fuss et al. 2015). There is also a clear distinction between the local stratigraphic levels concerning the preservation of organics. Large botanical remains are particularly poorly preserved in HAM5 and are mostly indicated by a light purple-coloured layer in the clay, whereas the actual organic matter is missing due to oxidisation and degradation processes. In HAM4, on the other hand, pieces of wood are three-dimensionally carbonized and pyritised, which makes preparation and conservation almost impossible (Fig. 5.2.1).

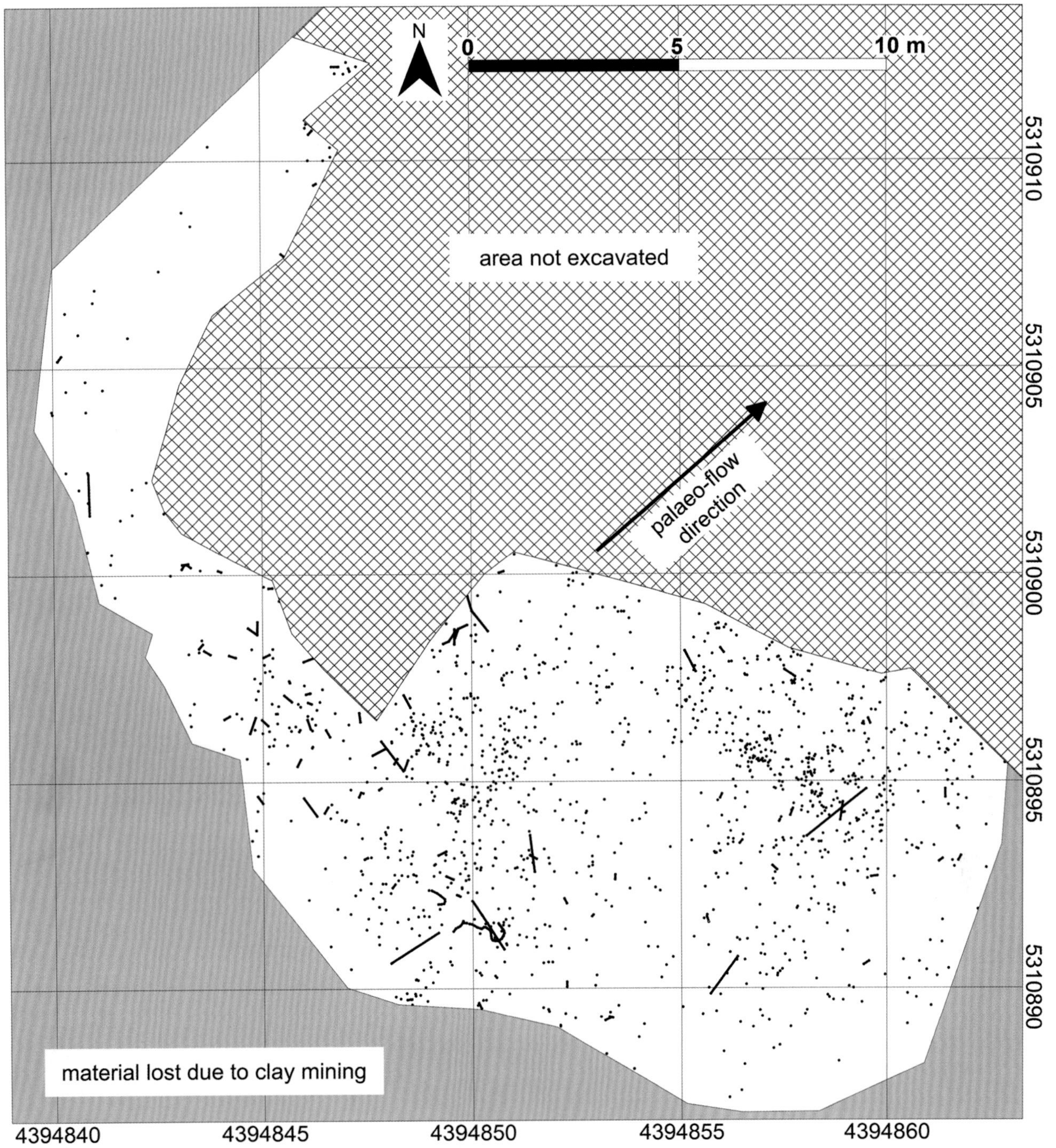

Figure 5.1.2.6. Detailed map of find distributions in the northernmost excavation area of the local stratigraphic level HAM4 (2022-2 to 2024) at the early Late Miocene Hammerschmiede site. Black dots mark excavated and measured isolated finds (mostly bones). Black lines represent elongated objects (the longest represent wood finds) including length and orientation. The black arrow indicates the presumed palaeo-flow direction of the HAM4 river and refers primarily to the orientation of the elongated objects found. The grey area has been lost due to clay mining, and the shaded area includes unexcavated areas. Coordinates correspond to Gauss-Krüger Zone 4 grid in metres. Grid spacing equals 5 metres.

Leaf fossils are extremely rare in both layers and mostly only consist of badly preserved impressions. In the HAM4 area reworked fragments of calcareous and pyritised root tubules (rhizocretions) can be found (Fig. 5.2.2), whereas Kirscher et al. (2016) also reported on in situ root tubules. Particularly well preserved in both levels are charcoal pieces from micro charcoal to up to 65 mm in size, which are frequently found especially in HAM4 (Fig. 5.2.3) and are evidence of vegetation fires in the catchment area or the vicinity of the HAM4 river (see section 5.7 below). The freshwater pearl mussel is one of the most abundant finds within the HAM4 channel fill. Most of the shells comprise both valves, slightly or completely opened and internally filled with surrounding fine sand, suggesting a post-mortem relocation. A few exceptions of found *Margaritifera*

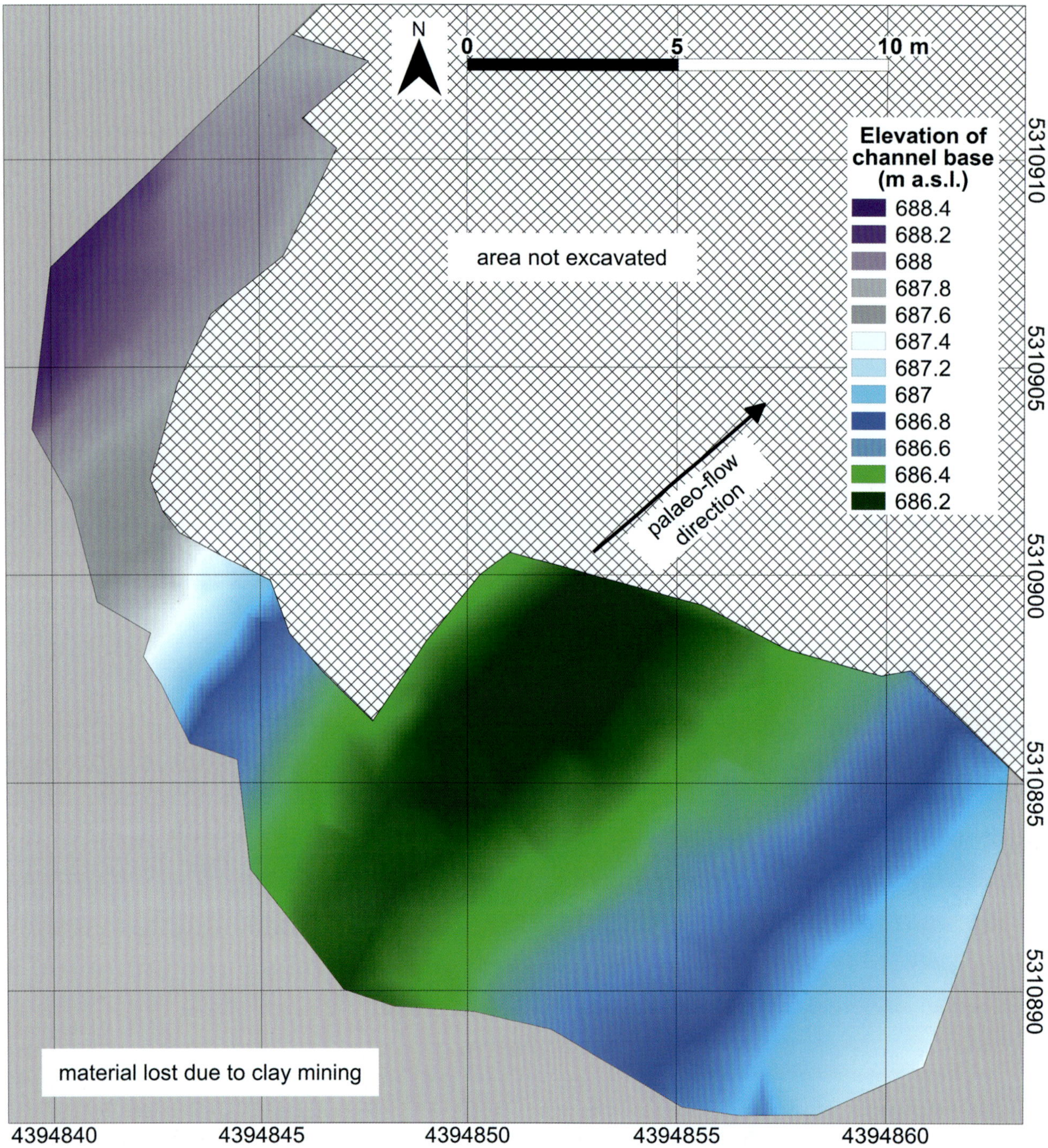

Figure 5.1.2.7. Relief map of the fossil river channel base elevation in the northernmost excavation area of the local stratigraphic level HAM4 (2022-2 to 2024) at the early Late Miocene Hammerschmiede site. The map shows the maximum erosion depth of the HAM4 channel based on measured points selected depending on the basement surface texture. The reconstructed palaeo-flow direction (arrow) is based on the course of the deepest channel base regions (darker colours) and the dip of the channel margin (brighter colours), supplemented by the orientation of elongated objects found. The greyed area has been lost due to clay mining, and the shaded area includes unexcavated areas. Values for the altitude in metres above sea level. Coordinates correspond to Gauss-Krüger Zone 4 grid in metres. Grid spacing equals 5 metres.

(*P.*) *flabellata* have a sediment-free interior and show a strong shell compression (Fig. 5.2.4 a–b). The enclosed core consists of a brownish, several millimetres thick organic filling (Fig. 5.2.4 b–d) which probably corresponds to remains of the original in vivo organics of the mussel and might represent a kind of soft-tissue preservation. Since these specimens are mostly found in thicker silty fine sand layers, a very fast embedding is indicated, most probably of the living shell which may have been killed by too great embedding depth.

Component-supported channel lags comprising enriched pedogenic carbonate concretions, rolled greenish clay pebbles, and sometimes mass occurrence of terrestrial gastropods (Helicoidea / *Tropidomphalus* sp.) indicate strong rearrangement and sorting processes (Fig. 5.2.5 a–b). The terrestrial gastropods from HAM4

Figure 5.2.1. Wood finds from the early Late Miocene Hammerschmiede deposits and the local stratigraphic level HAM4. (a) and (b) show typical finds of mostly elongated branchless wood that are used for palaeo-flow direction analyses by measuring the longitudinal orientation within the fluvial channel. Case (a) was documented in July 2019, and case (b) in August 2023. Reddish or brownish coatings (a) indicate a high content of iron sulphides that start to disintegrate after discovery. The length of the wooden piece in (a) is approximately 90 cm. The visible metre scale bar at the bottom provides a size reference. Scale bar in (b) equals 10 cm.

mostly show a greenish clay filling, while the main HAM4 components consist of greyish silts and fine sands (Fig. 5.2.5 c). This indicates an input of snails possibly washed in during rainfall from the surrounding terrestrial area of the HAM4 river. As gastropods act as sediment traps, they catch green sediments of the river peripheral flood plain. Alternatively, the terrestrial gastropods could be redeposited secondarily, presumably from the green clays into which HAM4 is cut. The same applies to the pedogenic carbonate concretions and the green mud clasts from the channel lag.

Bone and tooth preservation are very good in both HAM5 and HAM4 layers. The preservation of osteological matter nevertheless is very heterogeneous. Different taphonomic processes and pathways seem to be responsible for this. In many cases bones are perfectly preserved, showing a medullary cavity preserved as a void, sometimes with pyrite crystallisations (HAM4). Especially in HAM4, these bones also retain three dimensions, as the surrounding sediment (silt and fine sand) is hardly compressible and stabilises the bone. In the clay-rich layers of HAM5, this stabilisation is missing, as the clay compresses and drains under load. Several bones show significantly poorer preservation, indicating different degrees of transport distance and strength. Some bones are also damaged by biogenic processes (see section 5.9: Biogenic bone modification and possible producer). Some finds show surface corrosion, which could be abiogenic chemical or biogenic (e. g., algae, bacteria) in nature. Root traces indicate post-depositional rooting, which would have taken place before further redepositional processes. The bones of ontogenetically older and individually more ossified animals are often better preserved than those of young animals, which were probably still quite porous when embedded. Some finds show polishing and rolling of broken edges, indicating greater transport distances for these. With few exceptions of partially articulated finds and narrower strewnfields of single individuals, the innumerable loose individual bone and tooth finds indicate transport and sorting processes in the river course.

Figure 5.2.2. Reworked root tubule fragments from the early Late Miocene Hammerschmiede deposits of the local stratigraphic level HAM4 (2019). (a) fragment of a calcified root tubule in lateral (a1) and cross-section (a2) view. (b) Pyritized fragment of a root tubule in lateral (a1) and cross-section (a2) view. (a) GPIT/PL/21805; (b) GPIT/PL/21841. Scale bar equals 10 mm.

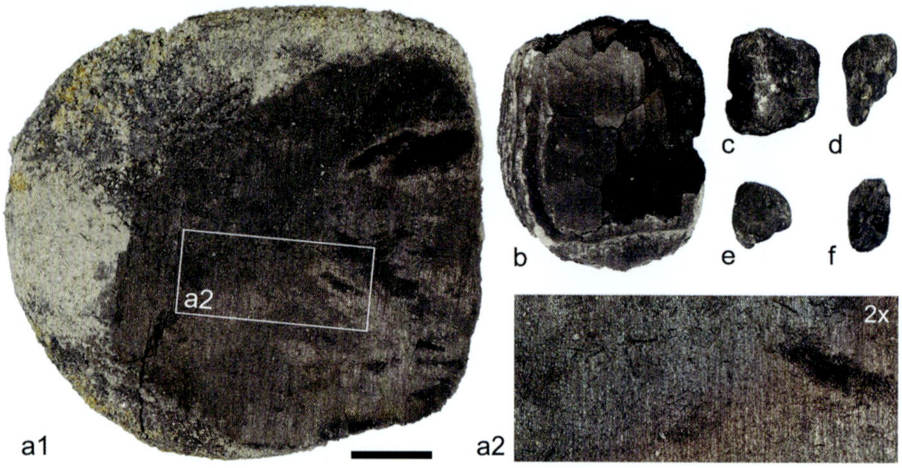

Figure 5.2.3. Charcoal specimens of different sizes from the early Late Miocene Hammerschmiede deposits and the local stratigraphic level HAM4 (a–f). Most specimens are rolled and compressed but show well-preserved wood cells (a2). Figure (a) represents the so far largest specimen of charcoal found at the HAM4 layer. The majority of charcoal specimens show smaller sizes (c–f). (a) GPIT/PL/21839, (b) GPIT/PL/21837; (c–f) SNSB-BSPG 2020 XCIV-7466. Scale bar equals 10 mm (a1, b–f) or 5 mm (a2).

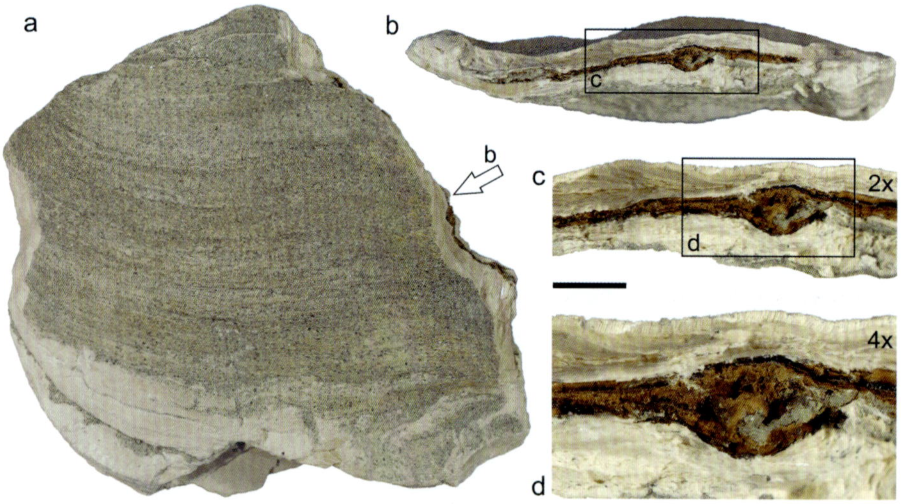

Figure 5.2.4. Preservation of organic matter within a freshwater pearl mussel (*Margaritifera flabellata*) from the early Late Miocene Hammerschmiede deposits and the local stratigraphic level HAM4. The shell shows a sand-free internal filling and strong compression (a–b). In the interior, a brownish mass is found (b–d), which probably corresponds to the remains of the original in vivo organics of the mussel and might represent soft-tissue preservation. (a–d) GPIT/BI/05904. Scale bar equals 10 mm (a–b), 5 mm (c) or 2.5 mm (d).

Figure 5.2.5. Terrestrial gastropods from the early Late Miocene Hammerschmiede deposits and the local stratigraphic level HAM4. (a) In situ, sample of a component-supported channel lag from the HAM4 channel deposits consisting of pedogenic carbonate concretions, rolled mud pebbles, and terrestrial gastropods (Helicoidea/*Tropidomphalus* sp.). (b) Completely preserved terrestrial gastropod (*Tropidomphalus* sp.) from HAM4. (c) Terrestrial gastropod (*Tropidomphalus* sp.) from HAM4 with partly broken and missing shell revealing a greenish mud filling (white arrow) while coarser sediments adhere to the outer shell. (a) GPIT/GA/05239; (b) GPIT/GA/05228; (c) GPIT/GA/05227. Scale bar equals 20 mm (a) or 10 mm (b–c).

5.3 Spatial distribution of finds (large scale)

The majority of bone finds from Hammerschmiede originate from sediment infill within incised channel courses. Due to the preferential accumulation and preservation of bones within these channel deposits, excavations have primarily focused on these stratigraphic units. The distribution of finds within the documented excavation areas, particularly in channel lag and lower point bar deposits, is largely governed by the hydrodynamic properties of the palaeocurrent, as well as the geological and geomorphological characteristics of the underlying bedrock. These processes result in a highly heterogeneous spatial distribution of fossils, rather than a uniform dispersion across the excavation area (Behrensmeyer 1988).

Hammerschmiede channels exhibit significant variation in thickness, lateral extent, and spatial positioning of fossil-rich layers, as well as in the density of embedded specimens (see Figs 5.1.1.2, 5.1.2.2, 5.1.2.4, 5.1.2.6). Additionally, changes in water discharge, shifts in flow direction due to meandering, and variations in channel base morphology – such as flat vs. undulating surfaces or the presence of pool-and-riffle structures – directly influence sedimentation dynamics. The physical composition of the channel base (e. g., muddy vs. consolidated substrates, and varying degrees of erodibility) further modulates depositional patterns, resulting in distinct sedimentary zones within the channel system.

Find densities within the channel deposits exhibit substantial variability, ranging from fewer than five to over fifty specimens per square metre nearby. The orientation of elongated objects, such as long bones and wood fragments, is influenced by the meandering nature of the palaeoriver, aligning with prevailing flow directions. This meandering process plays a crucial role in the accumulation of fossils within meander bends, where inclined, arcuate sediment packages form along the point bar as lateral and downstream migration of the meander progresses (Figs 3.2.4, 3.2.5). The sequential lateral shifting of the meander results in spatially heterogeneous grading patterns, typically characterised by a basal concentration of specimens with a gradual decrease in fossil density upwards within the sediment package.

In addition to inclined point bar deposits, channel lags – coarse-grained deposits accumulating parallel to the channel bed – also contribute to fossil preservation. Within these lag deposits and associated inclined point bar accumulations, disarticulated bones dominate, representing reworked and hydrodynamically enriched material. Closely associated or anatomically related specimens (i. e., articulated or semi-articulated finds) tend to be positioned at the uppermost parts of these deposits, often forming the final stratigraphic horizon before complete burial. In contrast, fossil occurrences within the more rapidly deposited fine-grained channel fills are rare, yet when present, they are often exceptionally well-preserved. This observation aligns with previous taphonomic findings by Behrensmeyer (1988), which emphasize the high degree of reworking in channel lag deposits and the superior preservation potential of material within finer-grained, rapidly buried contexts.

Beyond a depth of approximately 1–1.5 metres from the channel base, fossil abundance declines markedly, delineating the transition from the lower bar unit to the upper bar unit within the channel-fill sequence. Due to the substantial variability in channel depth between central and marginal areas, the absolute thickness of the lower bar unit is highly variable. While gastropod remains are ubiquitous in the clayey sediments across the entire clay pit area, large vertebrate fossils are notably absent outside the confines of the channel structures.

As a further indication of strong rearrangement processes, large- and small-scale sorting of objects can be detected. Only in the large-scale analysis, does it become clear that in the HAM4 certain fossil types are more frequent and accumulate in some areas while they are rare elsewhere. For example, areas with clustered mass accumulations of freshwater pearl mussels (*Margaritifera flabellata*) indicate sorting by density and object shape properties (see Fig. 5.5.2, below). Such processes can also be observed in HAM5. Here in the excavation area of 2018 there was a pool-like depression in the western lateral channel base, which accumulated material of greater density (carbonate concretions and especially mammalian teeth). The depression in the base with slight edge undercuts rather indicates higher currents and possibly whirlpool-like circulating water, which only allowed particularly dense material to retain here. More than 50 teeth and tooth fragments were found within less than two square metres, while the surrounding area supplied not even close to such a high number of teeth.

5.4 Spatial distribution of finds (small scale)

The find layers HAM5 and HAM4 provide next to a majority of small and medium-sized bones and fragments only a small quantity of larger bones. So far there is no find of a fully articulated skeleton. This already indicates a strong mixing, reworking, and transport of bones presumably also over larger distances. In addition, bones with excellent surface preservation can be found directly next to finds that have been crushed, corroded, rolled off, fragmented, and also covered with root marks. The largest quantity of finds are disarticulated bones or bone fragments from small and medium-sized vertebrates. Objects are very heterogeneously distributed and conclusions about related individuals are rarely possible. Especially the large mammal finds are in many cases limited to small fragments of teeth and bones broken down into small pieces of cm range.

Anatomically connected units or anatomically slightly disarticulated but correctly arranged units are usually comprised of extremities (e. g., Fig. 5.4.1 a–b). Bones are often arranged in anatomically correct order, but joints are not in situ (Fig. 5.4.1 a–b). Findings of this kind indicate that only ligaments were present at the time of deposition, which indicates a rearrangement after decomposition in a dry environment. There is only one case of true articulation in mammals comprised by a spinal column of an unidentified large mammal from the local stratigraphic level HAM5 (Fig. 5.4.1 c), in which the cranium and extremities are missing. Furthermore, there is the nearly complete and articulated bird leg of the new goose species *Allgoviachen tortonica* (Aves, Anatidae) from the HAM4 (Mayr et al. 2022). Finally, there are three articulated turtle shell finds among the reptiles of HAM4 (Reptilia, Chelonia: adult *Chelydropsis* sp.; adult *Testudo* sp.; juvenile *Clemmydopsis* sp.).

In some cases, partial skeletons can be assumed and reconstructed by matching parameters of spatial find density and proximity and strewnfield orientation according to the reconstructed flow direction, as well as taxonomic and age-specific characteristics, while there is no bone duplication and thus no intermixing with other individuals of similar characters. These so-called strewnfields are sometimes very small (radius less than 2 metres), or can sometimes be reconstructed over many tens of metres (Böhme et al. 2019a, Mayr et al. 2020b, 2022). A separate treatment of selected case studies can be found in the taphonomy section (see section 5.8).

Particularly noteworthy is the frequency of anatomically arranged units, which are usually located within a heterogeneous mixture of isolated individual finds. In general, such finds are not very easy to identify, resulting in only a few known cases from HAM4. Especially the lateral overbank extension of HAM5 consists of several distinguishable and well-identified strewnfields, which, compared to other channel areas, exhibit less reworking, narrower dispersion, and minimal mixing with material transported from distant sources. Here, lower flow velocities or shorter time and rearrangement factors seem to have acted, which at the same time only affected the last of the three sedimentation phases of HAM5.

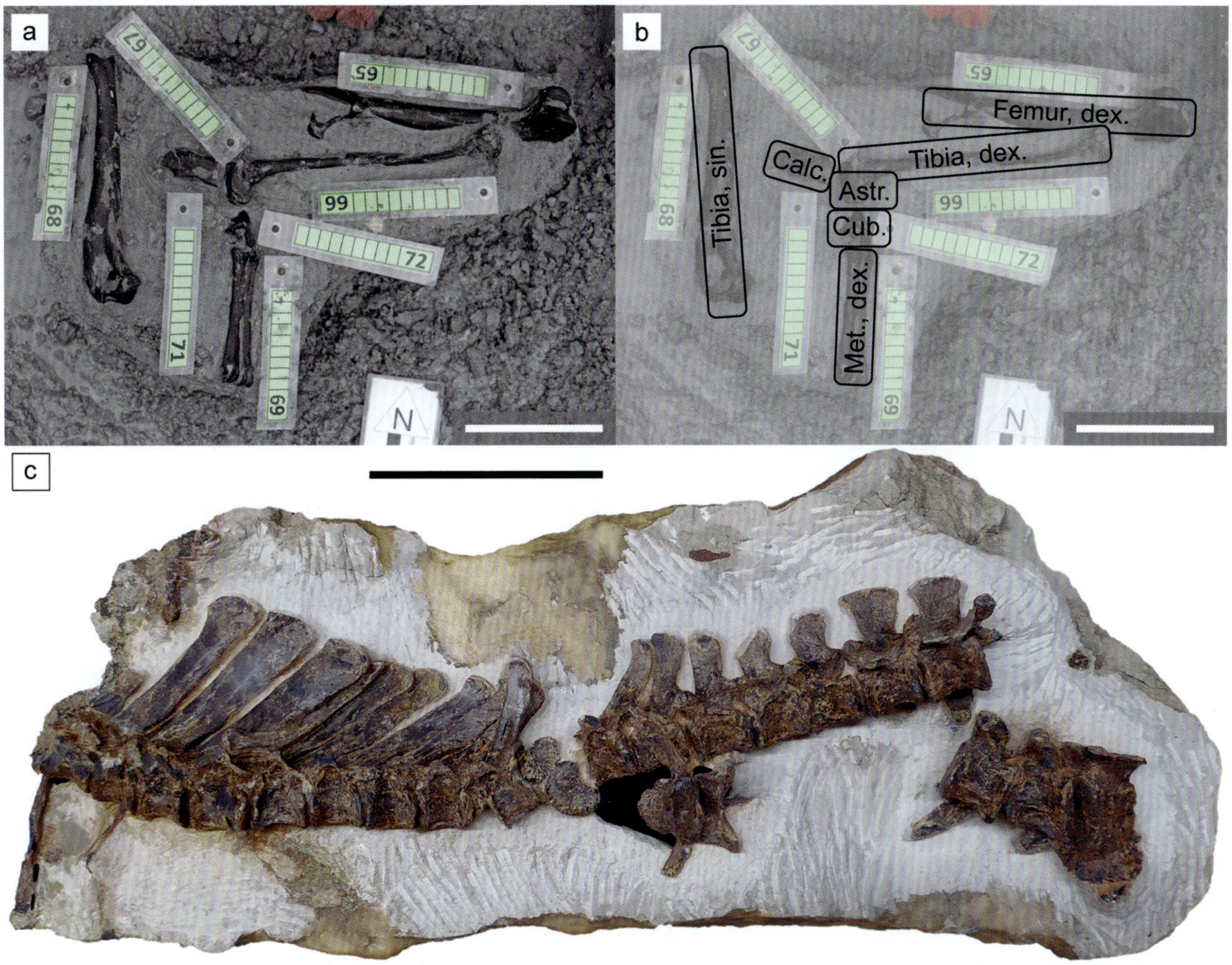

Figure 5.4.1. Examples of arranged or articulated mammalian skeletal finds from the early Late Miocene Hammerschmiede deposits and the local stratigraphic levels HAM5 and HAM4. (a–b) anatomically correctly arranged but not articulated hindlimb of the tragulid *Dorcatherium naui* (GPIT/MA/18000) very probably belonging to a single individual, found at the southernmost excavated section HAM4 2019-1. (a) Illustration of the find situation (August 2019). (b) Superimposed labeling of the bone determination. Abbreviations: Astr. – astragalus; Calc. – calcaneus; Cub. – cubonavicular; Met. – metatarsals; sin. – left side; dex. – right side (in anatomical position). The green number tags correspond to the field inventory number range 14XX. (c) The Spinal column of a yet unidentified large mammal from HAM5 is the only mammalian find from Hammerschmiede that exhibits true articulation (GPIT/MA/10999). Scale bars equal 10 cm.

5.5 Reconstruction of flow direction

Several indicators can be used to determine the direction of flow of the palaeochannels. On the one hand, the erosional channels of both HAM4 and HAM5 deposits themselves and the distribution of finds within them indicate the direction and course of the flowing waters. Furthermore, elongated objects and their orientation as well as the distribution patterns of originally connected skeletons of single individuals scattered by the water current (skeletal strewnfields) can be used to reconstruct the orientation and flow direction of the actual fossil river.

The sediments of the HAM4 channel are exposed at several places across the Hammerschmiede clay pit, which gives a good insight into the potential spatial channel course. The channel structure is exposed at the southern and northern edges of the pit (Fig. 5.1.1.1). In the central clay pit area, the HAM4 channel has been lost due to clay mining as well as larger sections of the eastern edge of this structure. The excavated and exposed areas of the HAM4 channel base clearly show that the HAM4 river does not flow linearly at all, but can rather be described by a slightly meandering, S-shaped course, crossing the approximately 200 m width clay pit. This means that palaeo-flow direction reconstructions can only provide spatially focused, local results and do not necessarily apply to the entire HAM4 area. Nevertheless, the main palaeo-flow direction corresponds to a course from SSW to NNE due to the channel orientation and outcrop observations.

The erosive channel base was analysed as a further indicator for a flow direction reconstruction. When digging for fossils, the channel fill is excavated and removed. Finally, the open channel base area (maximum erosion depth in the underlying clay) is measured with selected single points depending on the surface texture, and a relief map is created (Figs 5.1.2.3, 5.1.2.5, 5.1.2.7, 5.5.1). Based on the course of the deepest part of the channel and the dip of the channel margin, the course of the HAM4 palaeochannel can be reconstructed (Fig. 5.5.1). It can be seen that in all excavated areas a section of a channel structure is visible, consisting of higher elevated margins and a deeper central area (Fig. 5.5.1). The difference in elevation and the eroded relief of the channel base was examined across the excavation areas of HAM4 from the very south (HAM4 South) to the northernmost area (HAM4 North). A comparison of the lowest elevation points within the HAM4 channel reveals a northward dip of 2.1 metres between the southernmost and northernmost excavation sites over a distance of 135 metres. This corresponds to a gradient of 0.89° (1.55%). At first glance, such a relatively steep gradient over this length might suggest a primary flow direction towards the north.

However, a more detailed examination of the Hammerschmiede profile indicates a slight tectonic dip of the stratigraphy towards the north, with a measured gradient exceeding 0.5° but remaining below 1°. Specifically, the upper lignite seam exhibits a dip of approximately 0.85° (1.5%), while the HAM7 layer records a dip of 0.5° (0.8%) (see also section 3 and Fig. 3.0.4). This tectonic influence closely aligns with the gradient observed in the channel profile, suggesting that the apparent northward inclination of the HAM4 channel may not be a reliable indicator of the original palaeoflow direction. Instead, the observed dip likely represents a structural overprint, limiting its usefulness for definitive hydrodynamic interpretations.

At various locations in HAM4, mass accumulations of adult freshwater pearl mussels of the species *Margaritifera flabellata* (Unionida, Margaritiferidae) were observed. Due to their elongated shape (long axis of about 13 cm), an evaluation of the longitudinal axis orientation can be carried out to be used as a proxy for small-scale flow direction analyses. For this purpose, three mass accumulations of freshwater pearl mussels occurring in different areas of the HAM4 were exposed, and the individual shell orientations were documented (Fig. 5.5.2).

The long axis orientations were finally analysed in line direction histograms (Fig. 5.5.2 a3, b3, c3). It is particularly striking that the different survey areas yield different results, which is an indication of different directional changes of the palaeo-flow within the HAM4 channel and shows that such evaluations can always only apply to a specific small-scale area. All three evaluations show two slightly different main directional distributions of the shell longitudinal axes. Following e.g., Voorhies (1969), Behrensmeyer (1990), or Lyman (1994) one direction possibly corresponds to the orientation of mussels that are turned longitudinally into the current and the other complies with bivalves "rolling" transverse to the current (Fig. 5.5.2 a3, b3, c3). It is noticeable that the angles vary between the prevailing directions which depending on the locality, merge somewhat diffusely (Fig. 5.5.2 a3), show clear distinctions (Fig. 5.5.2 b3) or converge strongly (Fig. 5.5.2 c3). One reason for this could possibly be the different water flow velocities during times of deposition. If the velocity exceeds the threshold for "rolling" both directions tend to merge into a single one, representing the flow direction. This situation is depicted by our case from the southernmost channel exposure near its cut bank (Figs 5.5.2. c and 5.5.3. C). Lower flow velocities suggest the bimodal directions from the northern part of HAM4 South (Figs 5.5.2. b and 5.5.3. B), located distally of a meander loop.

Over slightly larger scales, several other long objects can be used in the HAM4 for flow direction analyses (Fig. 5.5.3). A particularly good proxy for larger scale flow direction analyses are the often more than one metre long pieces of wood that are regularly found in HAM4 lower point bar sediments. Equally helpful are the numerous long bones from larger mammals, which are also recorded with their longitudinal orientation.

Since the flow direction in a meandering river varies depending on the position along a meander loop, we analyse the orientation of longitudinal objects across multiple excavation plots in both the northern and southern halves of the area (HAM 4: 2017–2018, 2019-2, 2020–2024), as well as along an east–west division (HAM 4: 2019-1). As a result, the orientation trends indicated by line-direction histograms (Fig. 5.5.3) reflect quite well the presumed S-shaped HAM4 channel course from the geological observation (compare Fig. 5.5.1 and section 5.1.2) and the results from the orientation of unionid bivalves.

The proxies considered so far are better suited for assessing the possible orientation of the flow rather than determining its exact direction.

From a theoretical point of view, a general flow from alpine direction in the south would be reasonable, but with such a flat and even landscape, it can be assumed that rivers and streams meandered to a great extent, and thus an apparent reversal of flow direction could potentially also take place in river sections. To reconstruct the actual direction of water flow, bone strewnfields from distribution events of single vertebrate individuals can be used. The distribution of loose bones transported by currents follows specific patterns, which are determined by the properties of each bone type – specifically surface area, weight, and, consequently, bone density. Furthermore, the bone shape itself can act as an anchor or promote rolling (Voorhies 1969,

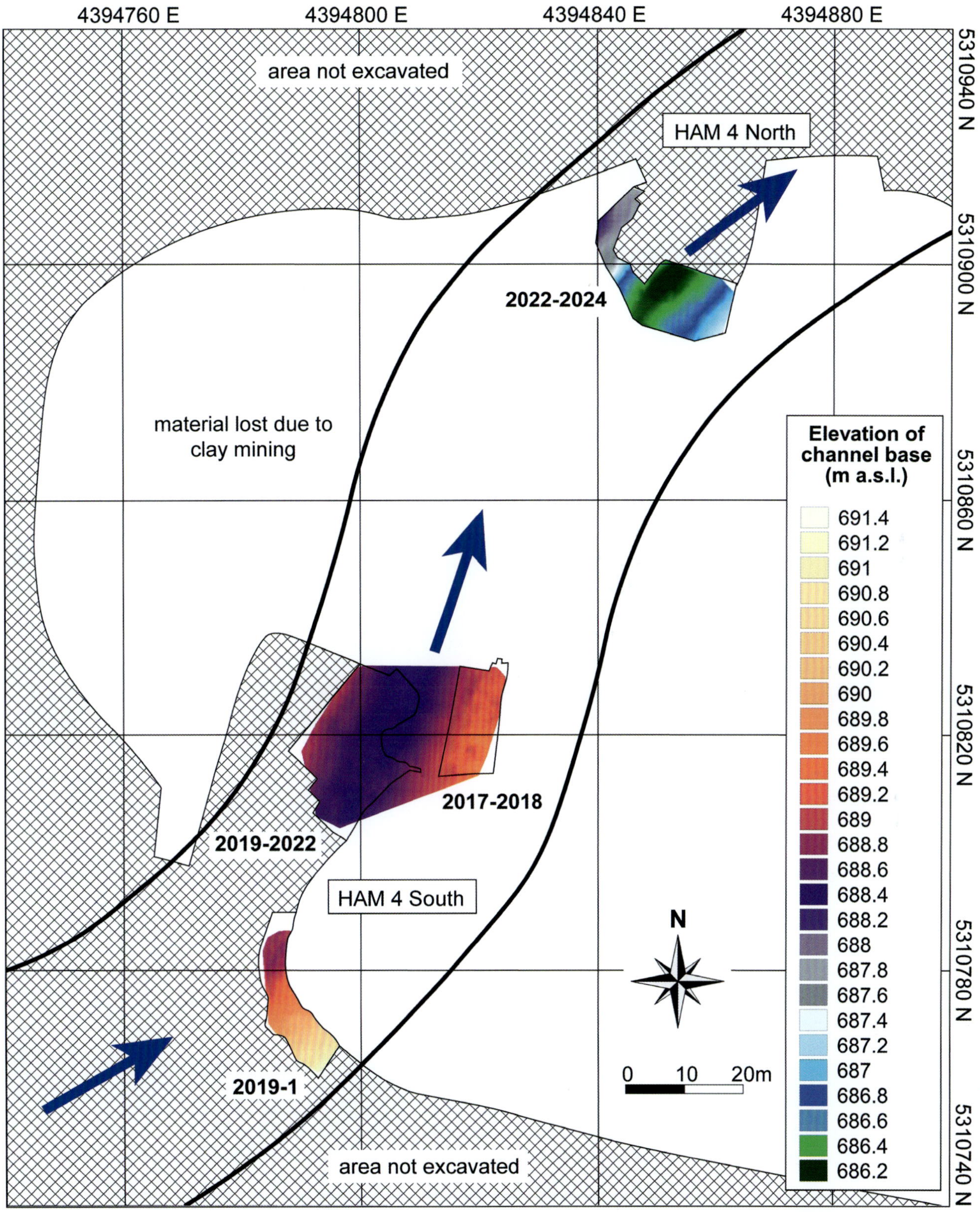

Figure 5.5.1. Relief map of the fossil river channel base elevation at the different excavation areas of the local stratigraphic level HAM4 at the early Late Miocene Hammerschmiede fossil site. The map shows the maximum erosion depth of the HAM4 channel based on measured points selected depending on the basement surface texture. Black polygons frame and combine neighbouring excavation areas HAM4 2017 to 2024. The section between the areas HAM4 2017–2018 and 2019–2022 has been lost due to clay mining and values for channel base elevation have been interpolated. Thick black lines indicate the assumed s-shaped meandering channel course based on outcrop observations of the channel structure. The reconstructed palaeo-flow direction (blue arrows) is based on the course of the deepest channel base regions (darker colours) and the dip of the channel margin (brighter colours), supplemented by the orientation of elongated objects found. The channel base dips to the north (colour/scale change). The observed inclination matches the values observed in all Hammerschmiede sediments (approximately 1.6% gradient to the north). Light grey colouring includes unexcavated areas and white areas have been lost due to clay mining. Coordinates correspond to Gauss-Krüger Zone 4 grid in metres. Grid spacing equals 40 metres.

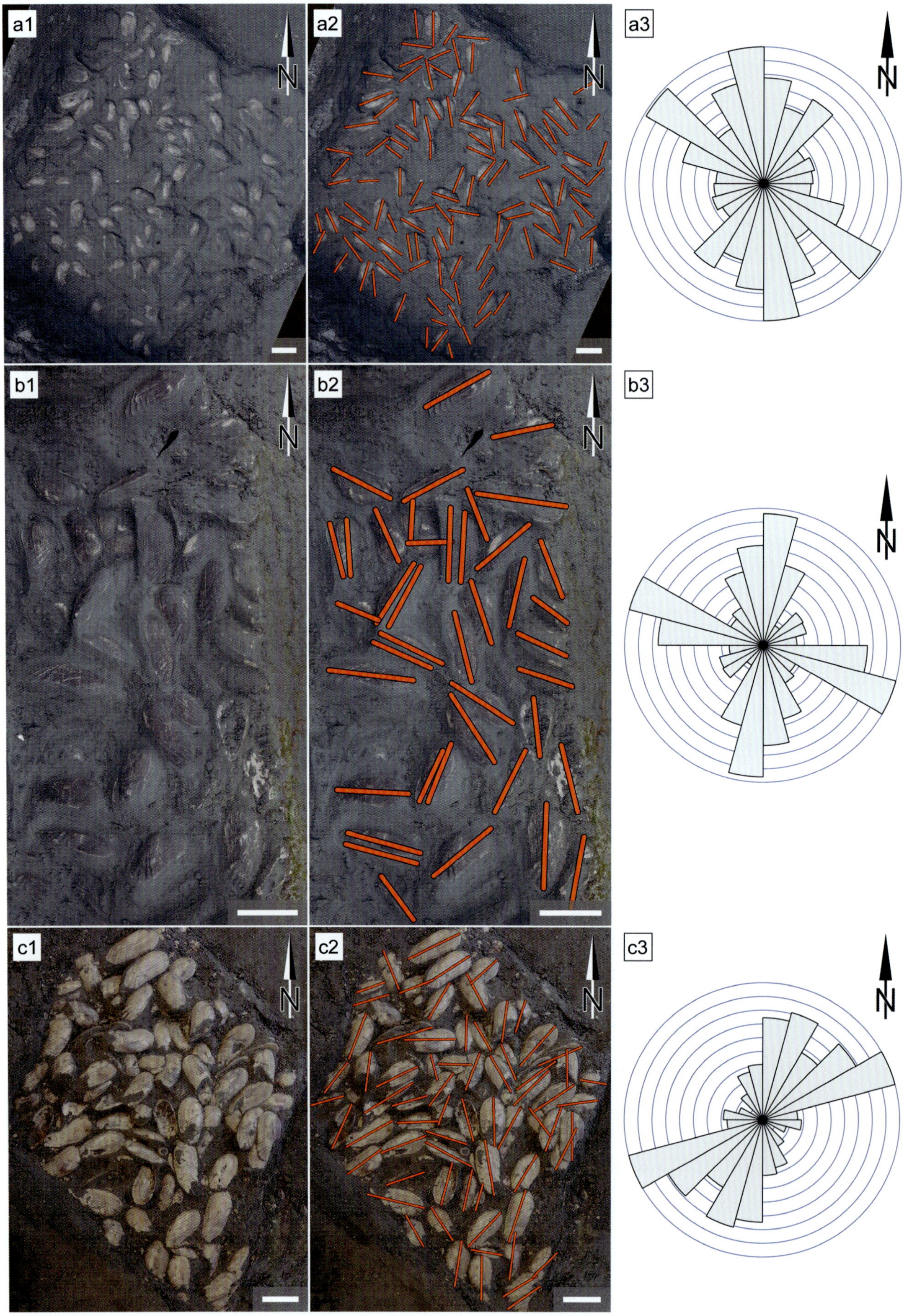
a1
a2
a3
b1
b2
b3
c1
c2
c3
N

◄ **Figure 5.5.2.** Mass accumulations of freshwater pearl mussels (*Margaritifera flabellata*) and longitudinal axis line direction histograms of the shells from the early Late Miocene Hammerschmiede deposits and the local stratigraphic level HAM4. Freshwater pearl mussel beds were carefully uncovered at three different locations (a1, b1, c1) based on the orientation of the long axes of the individual shells (a2, b2, c2), line direction histograms were created (a3, b3, c3), indicating the prevailing flow direction during sedimentation. (a) Shell deposits from the central channel (HAM4 2022-2; 119 orientations), (b) from the eastern channel margin at the central clay pit (HAM4 2017; 41 orientations), and (c) from the easternmost margin at the southernmost channel (HAM4 2019-1; 70 orientations). See Fig. 5.5.1 for positions of the excavation plots. All histograms show two main directions, which depending on the locality, merge somewhat diffusely (a3), show clear distinctions (b3), or converge strongly (c3). Scale bars equal 10 cm.

Behrensmeyer 1975, Hanson 1980, Behrensmeyer 1988). Particularly heavy and dense bones such as jaws and teeth, but also phalanges, should therefore lie more proximally, whereas vertebrae and ribs cover particularly distal strewnfield areas due to their less dense properties, which in some cases even extend to floating (Voorhies 1969, Behrensmeyer 1975, Hanson 1980, Behrensmeyer 1988). The longer a skeleton is exposed to currents, the exponentially longer the corresponding distribution area will be and the better the sorting (Voorhies 1969, Behrensmeyer 1975, Hanson 1980, Behrensmeyer 1988). In the present case, it appears that these arrangements found in the experimental tests (Voorhies 1969, Behrensmeyer 1975, 1988) can presumably be influenced by the physical properties of the river bottom and the sediment contained therein and may differ from the known patterns.

In the case of two *Dorcatherium naui* (Artiodactyla, Tragulidae) skeletons, it is evident that, in addition to the cranial elements and some limb bones, large portions of the ribs have remained in place, while the vertebrae, with a few exceptions, are missing (details see section 5.8). If a carcass decomposes and all bones disarticulate, the flat ribs, characterised by their large bone surface, can adhere to the ground. In contrast, vertebrae, with their more bulky and spiny three-dimensional shape, are more easily rolled out of position and transported away due to reduced contact with the ground. This process was facilitated by the fine-grained nature of the Hammerschmiede fluvial sediments, which were very muddy (silt and clay) and lacked coarse grains larger than medium sand (see section 3). However, most actuopalaeontological taphonomy studies have been conducted in rivers with quite different bottom conditions (Voorhies 1969, Behrensmeyer 1975). In addition, a carcass settled on the bottom that is exposed to the slowly increasing velocity of the flow is rarely considered. Thus, for example, certain areas may be exposed, excavated, and transported downstream while others may be trapped by the sediments encountered.

5.6 Syn- and postgenetic processes

5.6.1 Mineral neoformation and uranium incorporation

Postgenetic processes have affected the fossils of Hammerschmiede for millions of years. The bone colours have changed due to permineralisation and mostly show brownish to black, in the weathered oxic area reddish and yellowish colours. In most cases, these colours are probably due to the deposition of manganese- or iron-oxides and iron-hydroxides or iron-sulphides such as pyrite. Uncompressed bones (especially diaphyses of long bones or phalanges) from the HAM4 level sometimes show pyrite mineralisation in the bone cavity if the bone breaks when it is found and provides an insight (Fig. 5.6.1.1 a–b). In the HAM4 sands, iron sulphide (pyrite?) framboids of a few millimetres in diameter can be observed, which most probably prove an early post-depositional authigenic formation, as they are found isolated in the sediment and envelop the surrounding sand (Fig. 5.6.1.1 c).

Furthermore, bones from the local stratigraphic Hammerschmiede levels HAM5 and HAM4 emit slight radioactive radiations produced by incorporations of mostly uranium isotopes into the bones (Albrecht 2017). Hitherto unpublished investigations have shown that the bones of HAM5 exhibit a significantly higher enrichment of uranium isotopes than those of HAM4. The uranium enrichment was introduced over an early diagenetic period, e.g., via groundwater infiltration, and is likely due to the substitutional capability of uranyl (UO_2^{2+}) and calcium ions (Ca^{2+}) in the apatite of the bone. Uranium contents show a fluctuation of 0–3200 ppm (Albrecht 2017).

These partly high iron sulphides and uranium contents in HAM4 and HAM5 indicate a strong redox potential within the sediments. As the fluvial channels HAM5 and HAM4 are regarded as well-ventilated (rich in oxygen) by the observed faunal elements (e. g., salmonid fishes, freshwater pearl mussels, traces of ephemeropteran larvae), syn- and post-sedimentary processes remain causative for these anoxic mineral neoformations. In a study of redox reactions at the water-sediment interface in a modern Chinese river, Wang et al. (2020) provide decisive indications in this regard, which point primarily to the sediment compos-

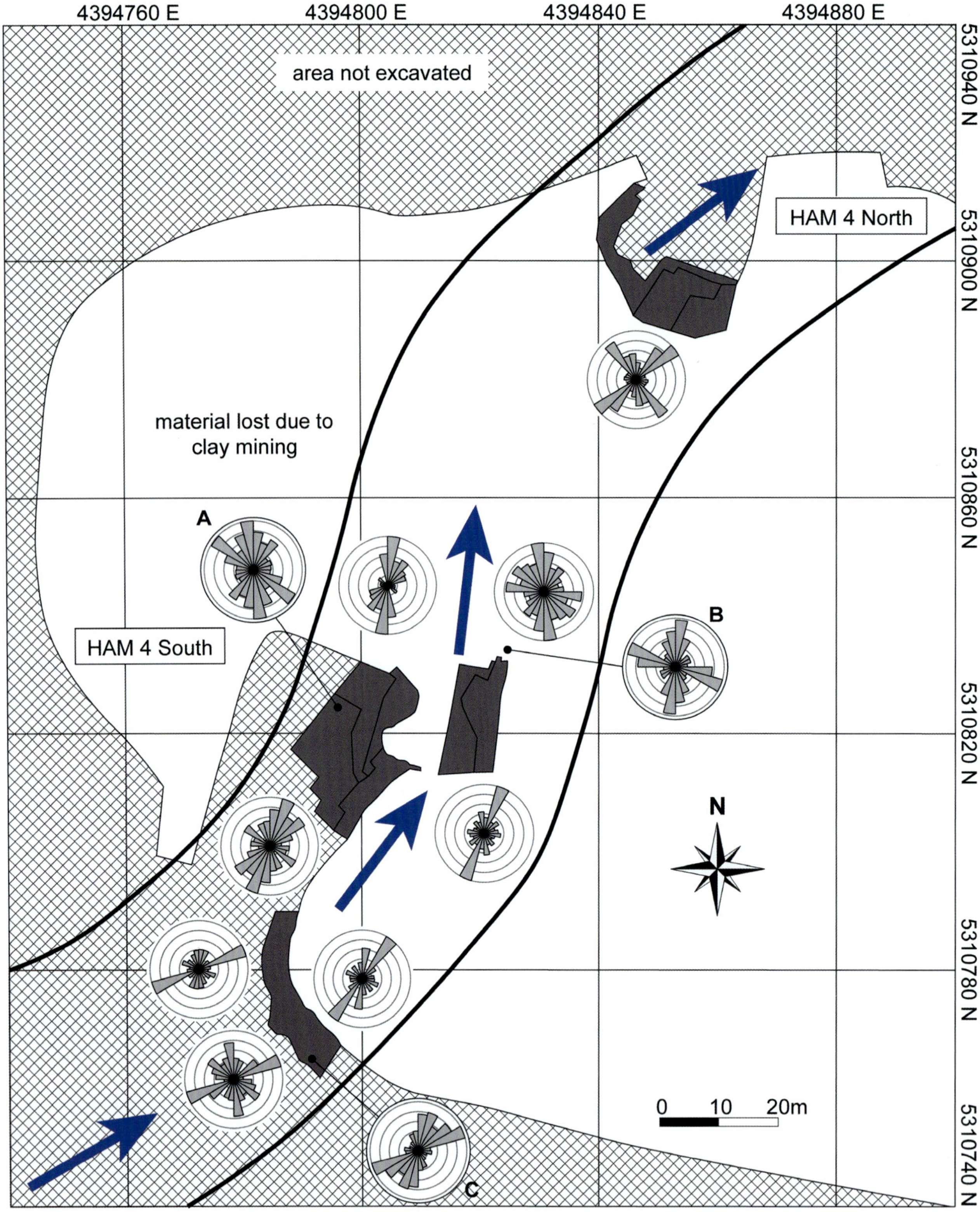

Figure 5.5.3. Line direction histograms from longitudinal objects (bones, wood, and bivalves) at different locations of the excavation areas of the early Late Miocene Hammerschmiede deposits and the local stratigraphic level HAM4. To be able to determine small-scale changes in the flow direction of the HAM4 channel, the excavation areas were analysed in several zones (approximate position of diagram near analysed area). The figure is supplemented by the line direction histograms of freshwater pearl mussels analysed in Fig. 5.5.2, localised within the HAM4 channel (A, B, C). Excavation areas in dark grey. Thick black lines indicate the assumed slightly S-shaped meandering channel course based on outcrop observations and blue arrows indicate the presumed and most likely direction (opposite direction also possible) of palaeo-flow based on the orientation of long objects resulting from the line direction histograms. Light grey indicates unexcavated areas and white areas have been lost due to clay mining. Coordinates correspond to Gauss-Krüger Zone 4 grid in metres. Grid spacing equals 40 metres.

Figure 5.6.1.1. Mineralisations from the early Late Miocene Hammerschmiede fossil site and the local stratigraphic level HAM4. (a–b) Authigenic mineral formation of iron sulphide (pyrite) within the cavities of vertebrate bones. (a) View into a femur diaphysis from the tragulid *Dorcatherium naui* with idiomorphic pyrite mineralisations. (b) Bone fragment with a view of the cavity surface with pyrite coverage. (c) Authigenic pyrite framboids from the silty fine-sand sediments of the HAM4 channel fill. Scale bar equals 10 mm.

ition in terms of grain size as the cause. In coarse-grained riverbed sediments, water is also transported through the already deposited sediments. To create anoxic redox conditions within the bottom sediments, this water exchange has to be limited (Wang et al. 2020). In particular, the high content of silt and clay ("siltation") reduces the permeability and thus the oxygen supply through and into the channel sediments and furthermore provides redox conditions for iron- or sulphate-reduction and thus also for the precipitation of uranium (Wang et al. 2020). The high silt (HAM4) and clay (HAM5) contents of the fluvial channel deposits are probably responsible for the reductive milieu in the sediments that was presumably established during and shortly after deposition.

5.6.2 Compaction

Several postgenetic processes are responsible for the preservability of the different types of material found at the Hammerschmiede site. One of the strongest postgenetic processes is the diagenetic compaction of the sediments mainly caused by the overburden loading (Fig. 5.6.2.1).

This compaction is mainly based on the reduction of pore space through grain reorientation and water expulsion (e. g., Ulmer-Scholle et al. 2015), especially of the clayey sediments, and is thus partly responsible for bone compression found in the Hammerschmiede deposits (Fig. 5.6.2.1).

The sediments of the local stratigraphic level HAM4 consist mainly of fine sands, which were already much more compacted when deposited and thus did not compress the bones much when further sediment load was deposited (Fig. 5.6.2.1 a). Consequently, the clay-rich sediments of HAM5 are particularly more affected by this type of compaction than the sandy HAM4 sediments due to the higher compressibility. The deposits of the HAM5 rivulet are more heterogeneous and especially a layer in the western lateral channel shows a basal, very fine-grained and greasy clay layer, which allowed a strong post-sedimentary compression, whereby some bones were literally rolled out and sometimes only the joints were preserved in their original three-dimensional shape (Fig. 5.6.2.1 b). Erosion (e. g., by glaciers during the Pleistocene) removed a considerably large amount of overburden. At present, there is no estimate of the actual lack of sedimentary overlay at this locality, but new approaches, especially concerning the rank of coalification (vitrinite reflectance), could clarify this in the future. Since the observed rank of coalification is placed around the lignite rank, the assumed sediment cover was probably not insignificant. The Miocene Hammerschmiede section is overlain by several metres thick glacial till deposits of Middle Pleistocene age (Riss glaciation, Kirscher et al. 2016). Consequently, loading by Pleistocene glaciers must be expected, which could also have contributed to the compaction.

Last but not least, there is weight-loading by the large construction machines without which it would not be possible to reach the fossil layers. Particularly when exposing the strata, it cannot always be ruled out that the heavy weight of the excavator's shears and compacts the underlying sediments, especially when there is no lateral sedimentary support due to the mining slopes. Especially the heavy compression and shearing of several bones found in a very greasy clay layer, only a few cm thick, in the lateral overbank extension of the HAM5 was most probably caused by the heavy machines (Fig. 5.6.2.1 b). In the same area of HAM5, in 2018 a slipping tailing pile also sheared a not inconsiderable area of the discovery layer. Finally, there are

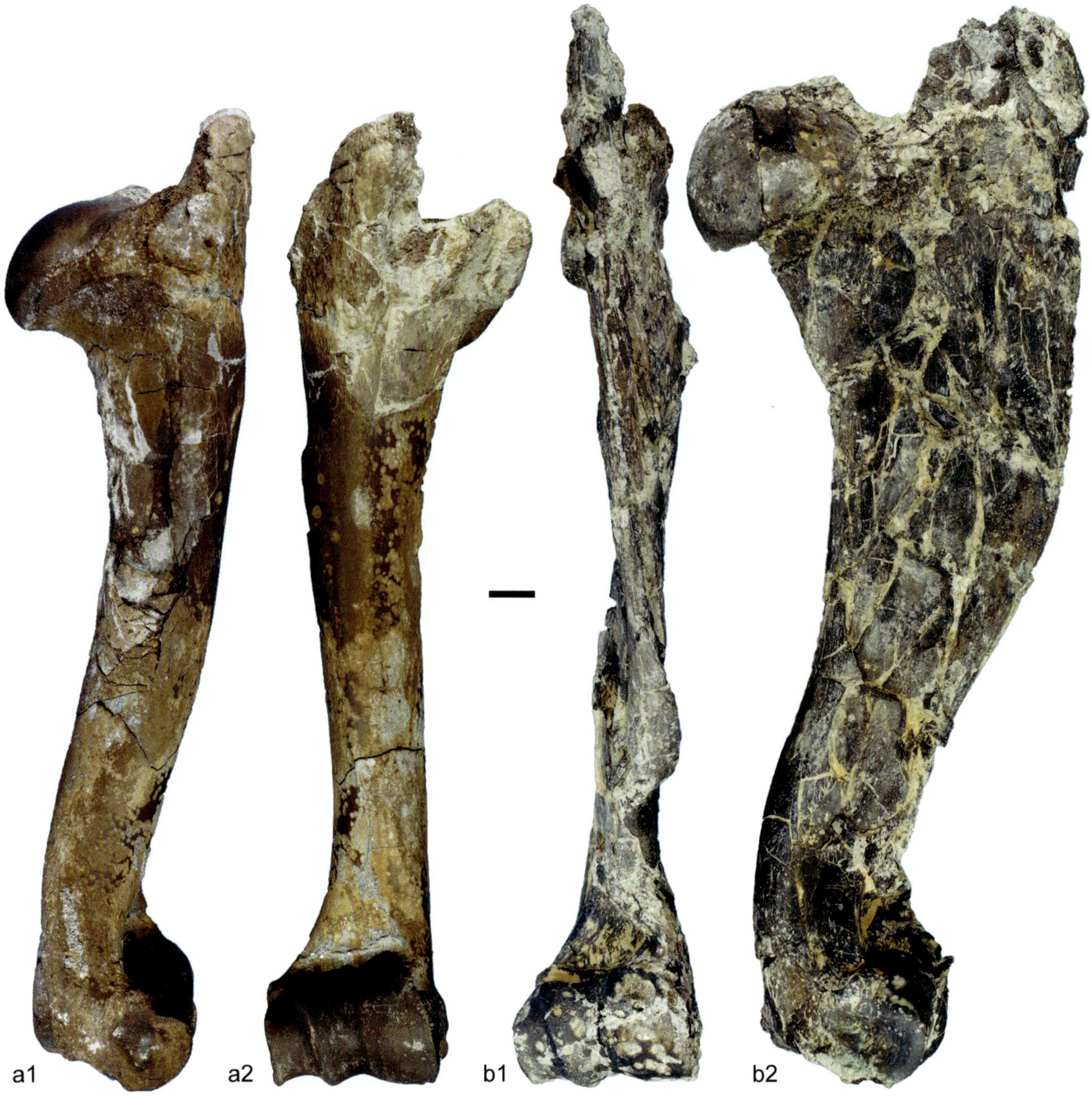

Figure 5.6.2.1. Different preservation conditions between the local stratigraphic levels HAM5 and HAM4 at the early Late Miocene Hammerschmiede site. (a–b) Two humeri of the antelope species *Miotragocerus monacensis* from the two different local stratigraphic levels HAM4 (a) and HAM5 (b) in lateral (a1, b2) and cranial (a2, b1) views. Note that the joints of the two bones shown indicate a humerus of nearly the same size. While the HAM4 humerus (a) is preserved almost intact in 3D, the HAM5 humerus diaphysis has undergone extreme medio-lateral compression resulting in significant lengthening (proximo-distal and cranio-caudal) while the distal joint itself is well-preserved. (a) SNSB-BSPG 2020 XCIV-7776; (b) SNSB-BSPG 2020 XCV-480. Scale bar equals 10 mm.

the workers and palaeontologists themselves, who of course may have caused damage to the finds in the form of mechanical fractures and displacements. The very sandy and silt-rich deposits of HAM4 show significantly less susceptibility to these contemporary pressures. Bones were recovered only a few centimetres below the excavator trackway, which left the internal cavity of the shaft intact. In most cases, any compressive damage that occurs affects the bottom sides of bones, which have collapsed. This is possibly caused by an historical ground failure of the surrounding sediments due to most probably very high lithostatic overburden pressures during diagenesis, as fragments are stabilised by mineralisation (Fig. 5.6.2.2).

Figure 5.6.2.2. Compressive damage caused by diagenetic overburden pressures at the early Late Miocene Hammerschmiede fossil site and the local stratigraphic level HAM4. (a–c) Radius of the bovid antelope *Miotragocerus monacensis* with, for the HAM4, typically impressed bottom side of the bone. (a) Anterior view of the radius, which pointed upwards at the discovery situation. (b) Posterior side of the radius, which in situ pointed downwards to the channel base. (c) Detailed view of the bottom side of the bone, which has been massively imprinted, while the top side remained intact. The indicated frame is only approximate, as (c) shows an oblique photographic view of the area. Scale bar equals 10 mm.

5.7 Microfossil taphonomy of HAM5 and HAM4

5.7.1 Quantity of fossil remains

To investigate the environmental characteristics of layers HAM5 and HAM4 we performed a taphonomic analysis of two test samples. The samples are composed of 1350 g (HAM5) and 1750 g (HAM4) of screen-washed and decalcified sediments. The HAM5 test-sample derived from the rescue excavation within the main channel in 2016. The HAM4 test-sample stems from the excavation plot 2020. Minimum and maximum mesh size of the sieves are 1.0 and 5.0 mm. All identifiable fossil remains have been picked out under a binocular, excluding only fragments of cortical and trabecular bones. We recovered in total 22,000 fossils from both samples (Tab. 3).

In general, the quantitative results are rather similar for both layers, with a dominance of fish fossils, phosphatic micro-coprolites and wood fragments, accounting for 67% and 79% of all remains in HAM5 and HAM4, respectively (Fig. 5.7.1.1 a–b). The taxonomic composition of the aquatic vertebrates is also quite similar, indicating a principally comparable riverine environment (Fig. 5.7.1.1 c–d). Both samples are nu-

Table 3. Quantitative results of a test sample analysis from the Hammerschmiede fossil site and the local stratigraphic levels HAM5 and HAM4. The samples consist of 1350 g (HAM5, main channel 2016) and 1750 g (HAM4, 2020) of screen-washed and decalcified sediments.

Taxonomic group of fossil remains		Type of fossil remains	HAM 5 (2016)	HAM 4 (2020)
Mammals	Rodentia	molars	489	160
		molar fragments	483	64
		incisors	617	193
	Eulipotyphla	dentition	167	81
	Lagomorpha	dentition	13	20
	small mammals indet.	postcranial	743	220
		toothless mandibles	11	0
	Ruminantia	tooth fragments	9	3
	Suina	tooth fragments	1	8
	Musteloidea	tooth fragments	4	0
		total number of mammalian remains	**2,537**	**749**

Table 3 continued on next page ▶

◄ *Table 3 continued*

Taxonomic group of fossil remains		Type of fossil remains	HAM 5 (2016)	HAM 4 (2020)
Reptiles	Chelonia	carapax fagments, claws	51	104
	Anguidae	osteoderms, vertebrae, skull fragments	164	169
	Serpentes	vertebrae, skull fragments, ribs	204	246
	Choristodera	vertebrae, skull fragments	8	14
	Scincomorpha	vertebrae, skull fragments	7	36
	Reptilia indet.	fragments	8	68
		total number of reptilian remains	**442**	**637**
Birds	indetermined	longbone fagments, claws	14	8
Amphibians	*Andrias scheuchzeri*	teeth	9	3
	Mioproteus sp.	vertebrae	659	423
		atlas	23	11
	Palaeoproteus sp.	vertebrae	19	11
		atlas	0	2
	Proteidae indet.	cranial bones	38	0
	Chelotriton sp.	vertebrae, cranial bones	2	85
	Salamandridae indet.	vertebrae, postcranial bones	4	6
	Latonia gigantea	cranial and postcranial bones	13	55
	Pelophylax sp.	cranial and postcranial bones	2	20
	Hyla sp.	postcranial bones	2	1
	Eopelobates sp.	cranial bones	0	1
	Anura indet.	vertebrae, cranial and postcranial bones	72	129
	Amphibia indet.	cranial and postcranial bones	0	43
		total number of amphibian remains	**843**	**790**
Fishes	*Leuciscus* sp.	pharyngeal teeth	1,670	1,319
	Tinca sp.	pharyngeal teeth	17	58
	Luciobarbus sp.	pharyngeal teeth, fin-spines	119	27
	Gobio sp.	pharyngeal teeth	0	1
	Cyprinidae	pharyngeal bones	225	431
		weberian vertebrae	292	522
		vertebrae and fin fragments, cranial bones	444	1,921
	Cobitis sp.	vertebrae, cranial bones	3	3
	Silurus sp.	proximal base of pectoral fin spine	466	425
		vertebrae, cranial bones	125	213
	Percidae div. sp.	vertebrae, fin fragments, cranial bones	82	94
	Gobius sp.	vertebrae, cranial bones	15	14
	Hucho sp.	teeth, vertebrae	6	17
	Esox sp.	teeth	2	1
		total number of fish remains	**3,466**	**5,046**
Coprolites	microcoprolites		3,684	1,494
Botanic remains	carbonized wood		933	1,060
	charred wood		44	198
	resin		6	4
	seeds		0	27
		total number of botanic remains	**983**	**1,289**
		total number of fossil remains	**11,969**	**10,013**

merically dominated by the catfish *Silurus*, followed by the chub *Leuciscus*. Water dogs (*Mioproteus*) are more common in HAM5, whereas barbels (*Luciobarbus*), tenches (*Tinca*), gudgeons (*Gobio*), gobies (*Gobius*), salmons (*Hucho*), pikes (*Esox*), perches (*Perca* spp.), and giant salamanders (*Andrias*) are comparatively rare or very rare (Fig. 5.7.1.1 c–d). However, because of the 1 to 5 mm sieve mesh size, these data did not reflect the actual abundance of the species in the ecosystem. The *Silurus* specimens in the test samples HAM5 and HAM4 all belong to juvenile or young catfishes with a body length between 6 and 30 cm (based on 1.5 mm, respectively 5 mm minimum and maximum mediolateral diameter of the proximal articulation of their pectoral spine bases; unpublished data by co-author MB), whereas the maximum adult length of *Silurus* in Hammerschmiede is estimated to 65 cm (based on the same method). In contrast, the cyprinid remains from both test samples belong to subadult individuals, whereas juvenile and adult cyprinids below 4 cm, respectively above 10–15 cm body length (vertebra length and pharyngeal bone fragments <1 mm and >5 mm) are not captured by our mesh-size.

The majority of cyprinid teeth show signs of resorption at their base, and pharyngeal bones with attached teeth are absent in both samples. However, the number of pharyngeal bones and Weberian vertebrae, relative to isolated teeth, is much higher in HAM4 (Tab. 3). According to the cyprinid tooth replacement method introduced by Böhme (2010), the tooth replacement quotient (TRQ) for *Leuciscus* in HAM5 is 1.07 (MNI 65, deduced from the numbers of isolated Weberian vertebrae), indicative for autochthonous cyprinid populations in perennial water bodies. In contrast, the TRQ for *Leuciscus* in HAM4 is only 0.45 (MNI 123). A TRQ <0.5 is, according to Böhme 2010 (p. 30), indicative of temporary waters, floodplains, or predatory accumulations. Since the sampled sediments derive from point

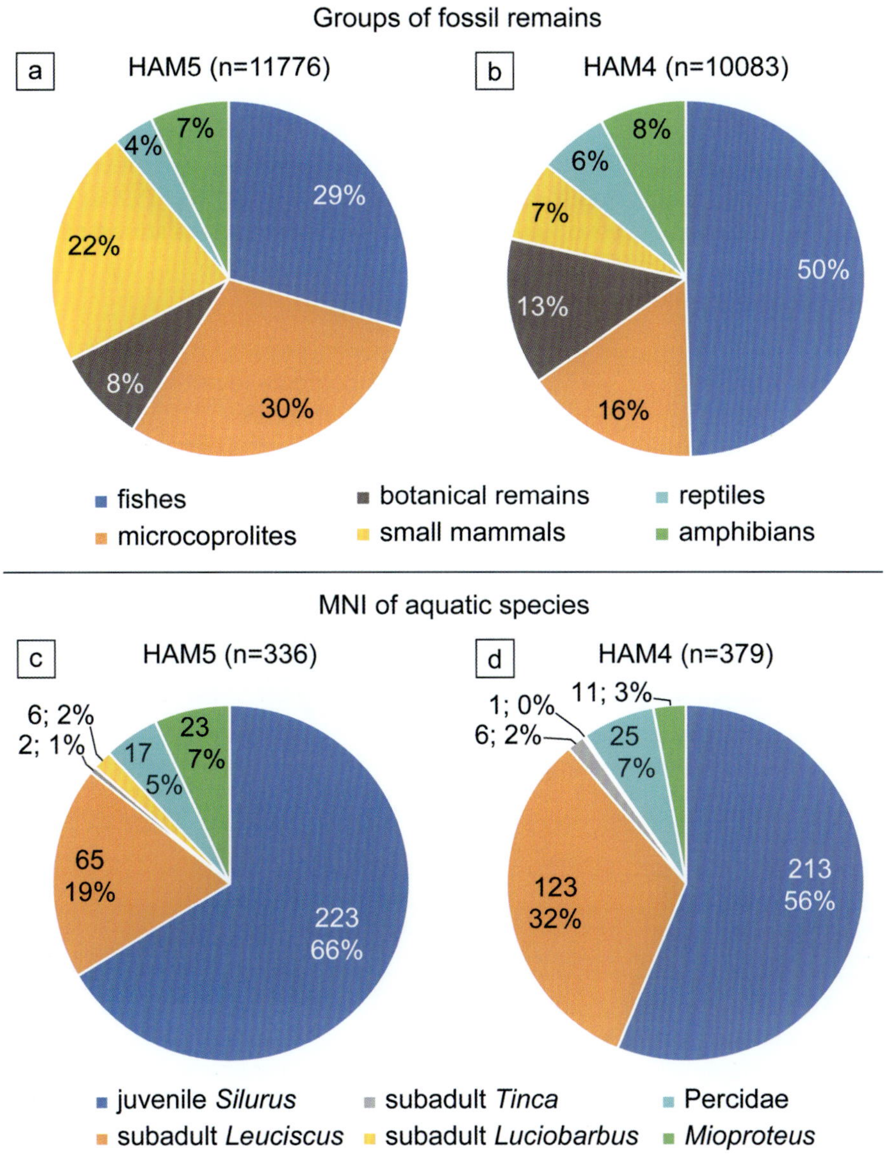

Figure 5.7.1.1. Relative frequency of different groups of fossil remains (a–b) and the minimal number of individuals (MNI) of aquatic species (c–d) from the Hammerschmiede fossil site and the local stratigraphic levels HAM5 and HAM4. Data refer to the quantitative analysis of samples from Tab. 3.

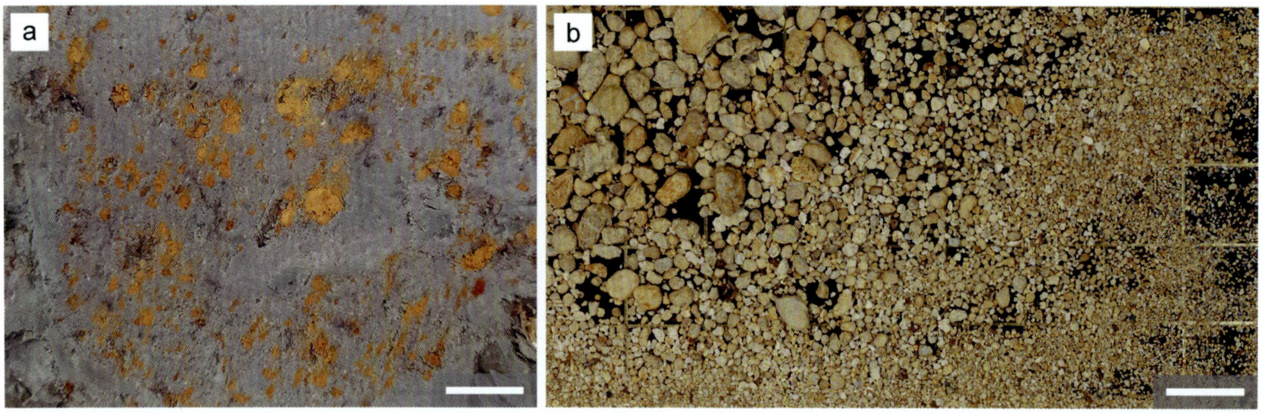

Figure 5.7.2.1. Microcoprolites from the early Late Miocene Hammerschmiede deposits and the local stratigraphic level HAM4. (a) Section through microcoprolite-enriched sediment, observed in the westernmost HAM4 area during the 2022 excavation season. The exceptionally high abundance of the orange-coloured microcoprolites is particularly striking, with hardly any other fossils macroscopically visible. (b) Overview of a small part of the screen-washed sample (mesh-size 0., 205 g dry weight) of this microcoprolite layer. The residuum (15%, 31,5 g dry weight) predominantly consists of phosphatic microcoprolites (0.2–) and fragments of fish bones. Scale bars equal 10 mm.

bars of a meandering river, an ephemeral water flow for HAM4 is not likely. The low TRQ rather suggests a stronger contribution of necromass supplied by predators to the HAM4 thanatocoenosis. Indeed, the number of fish predators is high in HAM4 (Tab. 3). They belong to birds (the dwarf cormorant Phalacrocoracidae indet. in the size of the extant *Microcarbo*, the darter *Anhinga,* and the kingfisher Alcedinae indet.; Mayr et al. 2020), mammals (three otters and two seals: *Paralutra jaegeri*, *Vishnuonyx neptuni*, *Lartetictis* cf. *dubia*, *Potamotherium* sp., Phocidae indet.; Kargopoulos et al. 2024b), but also to turtles (*Chelydropsis*, *Trionyx*) and fishes (*Silurus*, *Esox*, *Hucho*).

5.7.2 Microcoprolites and fish predation

The number of microcoprolites in HAM5 (30% of all fossils) significantly surpasses the amount recovered from HAM4 (16%) (Tab. 3). They are furthermore smaller in size (mean diameter 2 mm, mean weight 3.2 µg), relative to HAM4 (mean diameter usually 4–5 mm, mean weight 18.1 µg). The producers of these phosphatic coprolites are at present unknown, but the faeces of piscivorous birds (see above) seem to be a potential source. In any case, the less numerous and larger-sized microcoprolites in HAM4 indicate probably a sorting-out of larger size fractions on the point bars of the meandering river HAM4. The westernmost section of HAM4, excavated in 2022, revealed visually distinct layers significantly enriched in coprolites (Fig. 5.7.2.1 a). This coprolite-rich layer directly overlies the westernmost channel base, extending across several square metres within a clay-rich deposit only a few centimetres thick. Nowhere else in the Hammerschmiede locality such a high density of coprolites has been recorded.

We investigated a sample of this layer of accumulated coprolites by screen-washing 205 g (dry weight) of sediment using a sieve mesh size of 0.2 mm (Fig. 5.7.2.1 b). The results are presented in Table 4. 15% (31.5 g) of the sediment dry weight sample consists of phosphatic microcoprolites (0.2–8 mm) with an estimated number of >10,000. Many of them contain visible fishbone fragments. The remaining non-coprolite fossils (n=325) are nearly exclusively (99%) composed of fish remains. The size ranges of almost all fish fossils point to juvenile individuals with body lengths

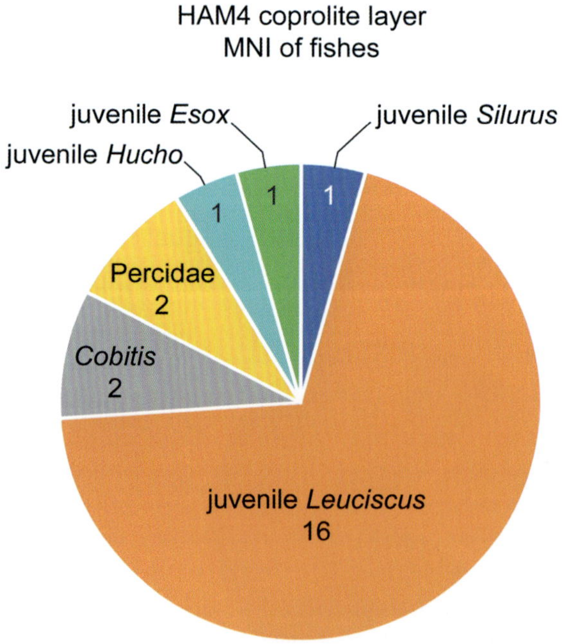

Figure 5.7.2.2. Minimum number of individuals (MNI) of fishes identified from a microcoprolite layer sample (see Tab. 4) at the early Late Miocene Hammerschmiede fossil site and the local stratigraphic level HAM4.

Table 4. Quantitative results of a microcoprolite layer from the Hammerschmiede fossil site and the local stratigraphic level HAM4.

Family	Species	Type of fossil remain	Count
Mammals	Rodentia sp. indet.	postcranial bones	2
Anurans	Anura sp. indet.	terminal phalanx	1
Fishes	*Leuciscus* sp.	pharyngeal teeth, resorbed base	1
		pharyngeal tooth germs	4
		basal broken pharyngeal teeth	16
	Tinca sp.	pharyngeal teeth	0
	Luciobarbus sp.	pharyngeal teeth	1
	Cyprinidae	pharyngeal bones	32
		weberian vertebrae	32
		vertebrae, fin fragments, cranial bones	194
	Cobitis sp.	cranial bones	3
	Silurus sp.	proximal base of pectoral fin spine	2
		vertebrae, cranial bones	1
	Percidae div. sp.	vertebrae, cranial bones	9
	Hucho sp.	teeth	23
		vertebrae	1
	Esox sp.	cranial fragments	2
		total number of fish remains	**321**
Botanic remains	seeds		1

>15 mm. This thanatocoenosis is dominated by *Leuciscus* (MNI=16), whereas mesopredators (Percidae n=2) and top predators (MNI=1 for *Silurus*, *Esox*, and *Hucho*) are rare (Fig. 5.7.2.2). The resulting TRQ for *Leuciscus* is 0.003 and the lowest ever recorded in fish thanatocoenosis of the North Alpine Foreland Basin (Böhme 2010). This result strongly suggests that all fish remains from this coprolite layer are derived from predation, and it is plausible to speculate that the fish predator and the microcoprolite producer may be identical.

5.7.3 Differences in the terrestrial palaeoenvironment between HAM5 and HAM4

Significant differences between the two layers were observed in the terrestrial component of the thanatocoenosis. HAM4 is characterised by a higher proportion of charcoal among the woody remains (16% versus 5% in HAM5) (Tab. 3; Fig. 5.7.3.1 a–b). Moreover, the representation of small mammal fossils within the vertebrate assemblage is markedly greater in HAM5 (22%) than in HAM4 (7%; Fig. 5.7.1.1 a–b). Small mammals (rodents and insectivores) account for 66% of all tetrapod fossils in HAM5, whereas amphibians, reptiles, and small mammal specimens are more evenly distributed in HAM4 (Fig. 5.7.3.1 c–d).

A comparison of the reptile suborders within Squamata – Anguimorpha (glass lizards), Scincomorpha (lizards), and Serpentes (snakes) reveals that the scincomorphs are more frequent in HAM4 (8%) than in HAM5 (2%) (Fig. 5.7.3.2 a–b). This difference is even more pronounced given that a single glass lizard individual (e. g., *Pseudopus*, *Ophisaurus*, *Anguis*) can produce hundreds of osteoderms, comparable to the high numbers of vertebral counts originating from a single snake individual. In contrast, scincomorphs are recovered mostly by gnathic remains, which suggests that they dominate the non-chelonian reptile fauna from HAM4 by the numbers of individuals. Furthermore, among small mammals, rodents dominate both assemblages, but lagomorphs (pikas) are more frequently encountered in HAM4 (8%) than in HAM5 (2%) (Fig. 5.7.3.2 c–d).

These differences suggest that the environment in the vicinity of the meandering riverine channel was a rather open forest in HAM4, with a higher frequency of wildfires and more open space occupied by lizards and pikas. In contrast, the environment of HAM5 suggests a small rivulet within the proximity of a rather closed habitat, accounting for the higher proportion of small mammals relative to other tetrapods (Fig. 5.7.3.1 c–d). This interpretation is confirmed by the abundance of certain rodent species, despite the fact that our samples are somewhat biased by the sieving-related lack

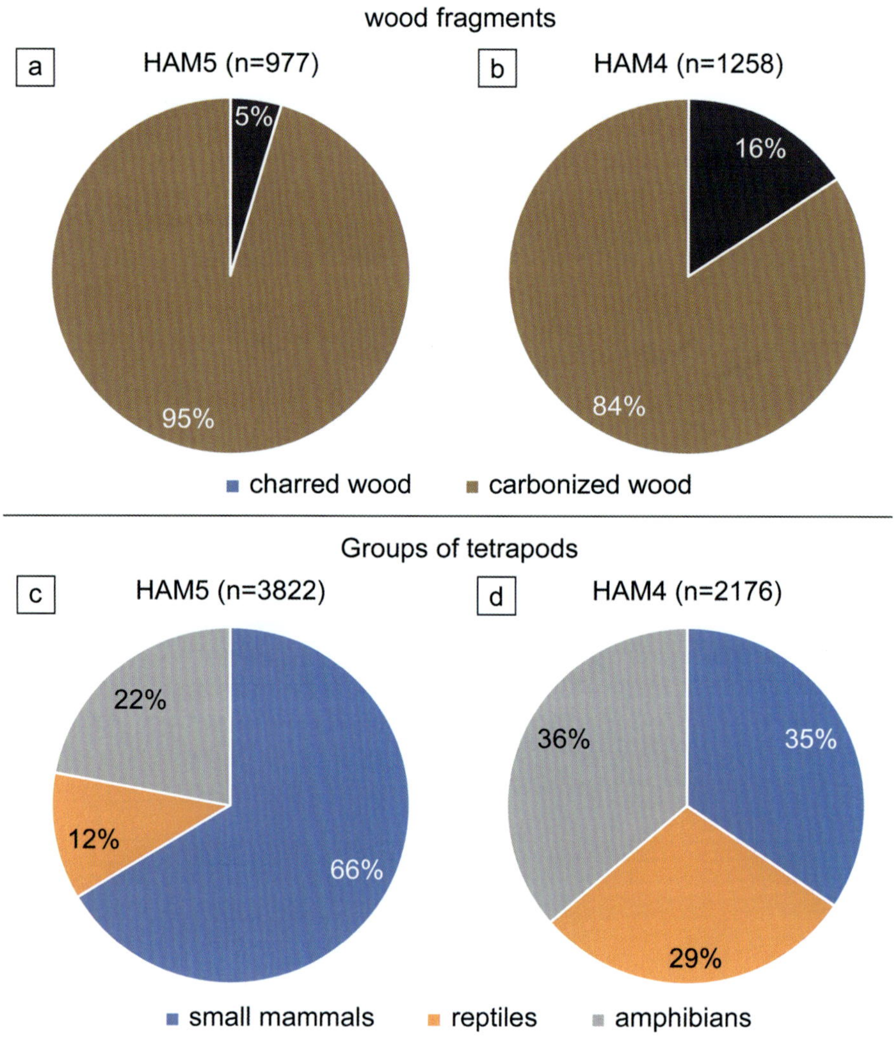

Figure 5.7.3.1. Relative frequency of wood fragments (a–b) and different tetrapod groups (c–d) from the early Late Miocene Hammerschmiede fossil site and the local stratigraphic levels HAM5 and HAM4. Data refer to the quantitative analysis of samples from Tab. 3.

of the smallest (Eomyidae with tooth size <1 mm) and the largest (*Steneofiber* with tooth size >5 mm) rodent remains (Tab. 5; Fig. 5.7.3.3 a–b).

The HAM4 rodent sample is dominated by *Microtocricetus molassicus*, a hypsodont microtoid hamster probably living in moist riparian habitats consuming reed leaves (Kretzoi & Fejfar 2005), and by the equally abundant *Csakvaromys bredai*, a marmotin squirrel (Tab. 5; Fig. 5.7.3.3 b). *Csakvaromys* (junior synonym *Spermophilinus*) was traditionally interpreted as a ground squirrel (de Bruijn 1999, Prieto et al. 2014), but Sinitsa et al. (2021) revealed that its cranial anatomy resembles a frugivore species with arboreal or semi-arboreal habit within the marmotin clade Sciurotamiina (rock squirrels). The only flying squirrel from the HAM4 test sample is the giant flying squirrel *Albanensia grimmi*, whereas the HAM5 sample yielded in addition the dwarf flying squirrel *Blackia miocaenica*, which is lacking in the HAM4 test sample (Tab. 5). The dominant rodent in HAM5 is again *M. molassicus*, followed by the anomalomyid *Anomalomys gaudryi* and the hamster *Democricetodon* nov. sp. (sensu Kälin & Engesser 2001, = *D.* cf. *freisingensis* in Kretzoi & Fejfar 2005) (Tab. 5; Fig. 5.7.3.3 a). Anomalomyids are interpreted as fossorial rodents digging with their front legs (Mein 1967, Bolliger 1999a, Nesin & Kovalchuk 2021). *Democricetodon*, which is rather rare in HAM4, is regarded by Prieto et al. (2014) as a potential forest dweller.

This leads us to interpret a more closed forest environment for HAM5, in contrast to a relatively open (riparian) forest for HAM4. This fits well with the results obtained from the Chalicotheridae (Kampouridis et al. 2024), which demonstrate the presence of a cursorial schizotheriine in HAM4, whereas HAM5 harbours the knuckle-walking chalicotheriine *Anisodon*. Furthermore, the presence of two syntopic hominid species in HAM5 (Böhme et al. 2019a, 2024) fits this scenario also quite well, although this interpretation has to be confirmed by palynologic studies in the future.

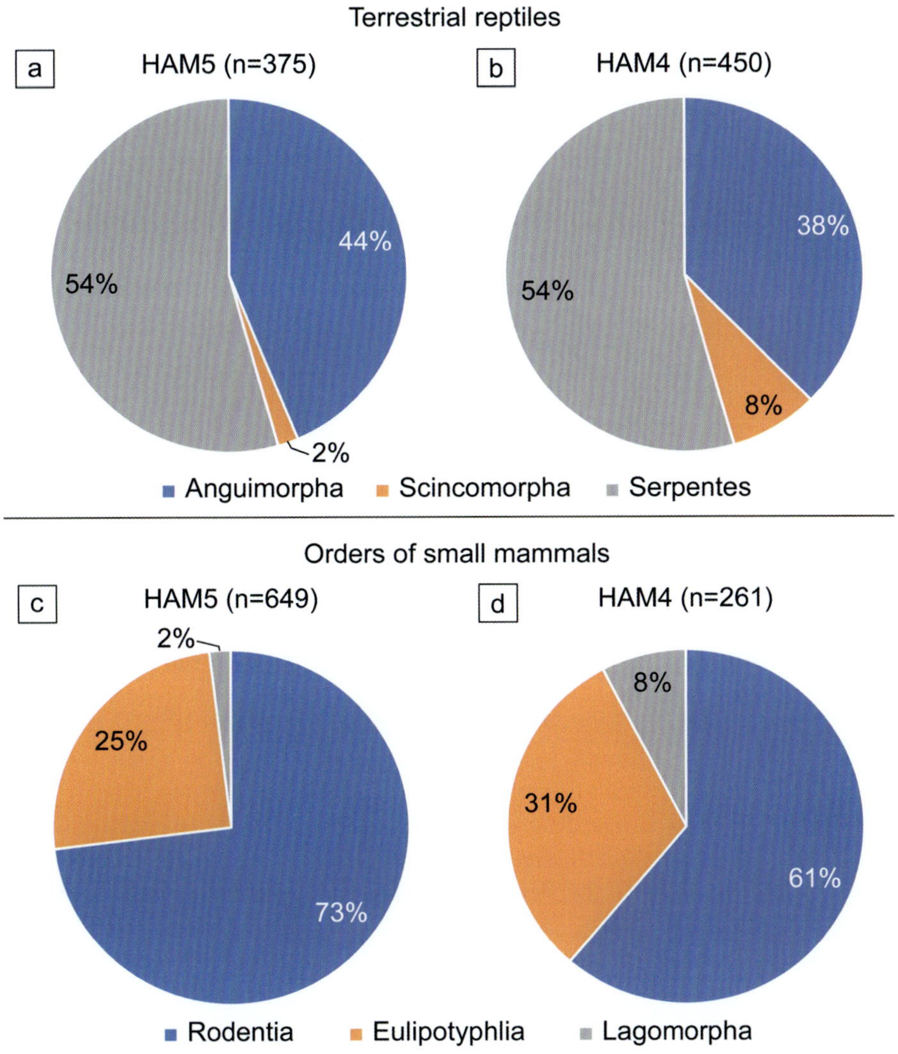

Figure 5.7.3.2. Relative frequency of terrestrial reptiles (a–b) and different orders of small mammals (c–d) from the early Late Miocene Hammerschmiede fossil site and the local stratigraphic levels HAM5 and HAM4. Data refer to the quantitative analysis of samples from Tab. 3.

Table 5. Relative frequency of certain rodent species (number of rodent molars) based on a quantitative analysis of samples from the Hammerschmiede fossil site and the local stratigraphic levels HAM5 and HAM4. The samples were composed of 1350 g (HAM5, main channel 2016) and 1750 g (HAM4, 2020) of screen-washed and decalcified sediments. Data on the presumed habitat requirements of the different rodent species are taken from [1]Prieto et al. 2014, [2]Kälin & Engesser 2001, [3]Hir 2005, [4]Kretzoi & Fejfar 2005, [5]Sinitsa et al. 2021, [6]Bolliger 1999a, [7]Lechner & Böhme 2023.

Family	**Species**	**HAM 5 (2016)**	**HAM 4 (2020)**	**Environment**
Cricetidae	*Megacricetodon minutus*	5	0	ubiquist? [1]
	Eumyarion latior	29	1	humid and forest? [1]
	Democricetodon nov. sp. [2]	58	5	forest dweller? [1]
	Collimys hiri	10	3	unknown [3]
	Microtocricetus molassicus	117	40	moist riparian or paludal environments [4]
Sciuridae	*Albanensia grimmi*	10	9	forest dweller [1]
	Blackia miocaenica	7	0	forest dweller? [1]
	Csakvaromys bredai	54	40	open habitat [1], arboreal to semi-arboreal [5]
Anomalomyidae	*Anomalomys gaudryi*	108	23	fossorial [6]
Gliridae	Gliridae div. sp.	38	10	forest dweller [1]
Castoridae	*Euroxenomys minutus*	53	29	semi-aquatic [7]
	total number of rodent molars	**489**	**160**	

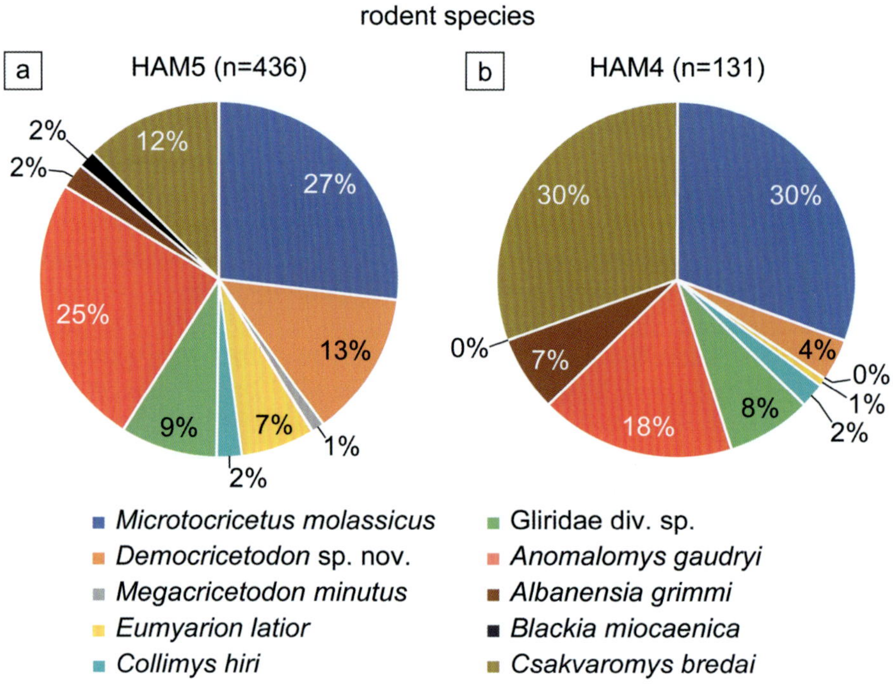

Figure 5.7.3.3. Relative frequency of selected rodent species (a–b) from the early Late Miocene Hammerschmiede fossil site and the local stratigraphic levels HAM5 and HAM4. Data refer to the quantitative analysis of samples from Tab. 3.

5.7.4 Quantification of vertebrate necromass

Based on our test samples described above and our excavated as well as field-washed finds we can roughly estimate the vertebrate necromass in lower point bar and channel lag sediments from HAM4 (Fig. 5.7.4.1).

For this purpose, all vertebrate fossils are grouped by size. The nano-size comprises specimens between 0.2 and 1 mm, the micro-size between 1 and 5 mm, the meso-size between 5 and 20 mm ('Rosie' finds), and the macro-finds comprise excavated fossils larger than 20 mm. The number of finds refers to one metric ton of unprocessed raw sediment. These rough estimates reveal that one metric ton of sediment yields over 1.6 million vertebrate fossils, the majority of them (99.76%) belong to the nano-size group (e.g., mostly juvenile fishes leftover by predators, see Tab. 4). However, it must be pointed out that this exceptionally large necromass volume in the nano-size group is primarily based on the evaluation of a specific microcoprolite-rich layer. Meso- and macro-scaling fossils recovered in the field by manual excavation and 'Rosie'-sieving (e.g., large mammals, giant salamanders, chelonians) comprise only 12 specimens per metric ton, whereas the micro-fraction (e.g., small mammals, amphibians, non-chelonian reptiles, juvenile and adult fishes) are estimated to about 3.900 specimens per metric ton. These numbers show that the majority of necromass belongs to fishes, which is an expected outcome in fluvial deposits but can be easily overlooked when only larger field finds are considered. Fishes may therefore play a crucial role in nutrition in the trophic net of the Hammerschmiede fluvial ecosystem, not only for piscivorous predators (see conclusion), but potentially also as 'fertilizer' for the terrestrial ecosystem by nutrient transfer via excrements.

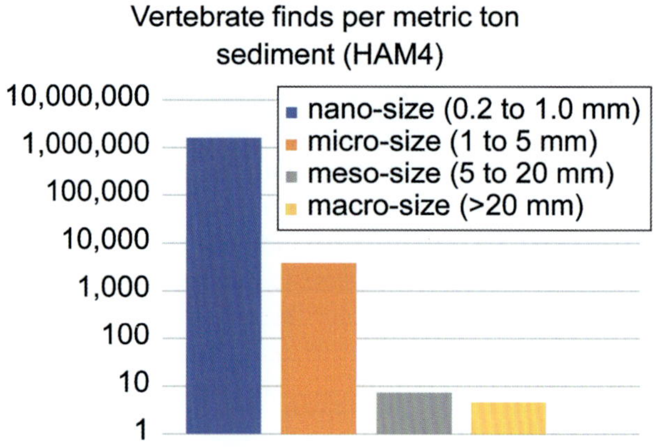

Figure 5.7.4.1. Quantitative necromass estimation for vertebrate finds per metric ton of sediment from the early Late Miocene Hammerschmiede fossil site and the local stratigraphic level HAM4. Data refer to a sediment sample of a particularly microcoprolite-rich layer at HAM4 (nano-size), the quantitative analysis of samples from Table 3 (micro-size), on-site readings from 'Rosie'-washed residues (meso-size), and measured finds during field campaigns (macro-size).

5.8 Case studies in large mammal biostratinomy

The deposits of the Hammerschmiede site include a large number of isolated finds, whose relationship to each other cannot be reconstructed at the moment. Among these objects, which were presumably repeatedly rearranged over long distances, there are nevertheless arrangements of bones that can be grouped as the same individual based on their taxonomic and age-specific identity as well as taphonomic parameters. The more excavation data are generated, the more case studies of partially articulated or disarticulated skeletal strewnfields become evident. Some of these examples have already been described in more detail in several studies (Böhme et al. 2019a, Mayr et al. 2020b, 2022, Konidaris et al. 2023). These include, as one of the most famous examples, the scattered remains of a male ape of the newly discovered ape species *Danuvius guggenmosi* at the site in addition to several remains of juveniles and females (strewnfield length 30 metres; Böhme et al. 2019a). Due to its taxonomic particularity, this example is also discussed again in more detail here. Disarticulated avian bones of a darter – *Anhinga pannonica* (strewnfield length 10 metres; Mayr et al. 2020b), the articulated leg of a new tree goose – *Allgoviachen tortonica* (articulated leg; Mayr et al. 2022), bones from the right wing region of a small dabbling duck – *Mioquerquedula* sp. (strewnfield length 10 metres; Mayr et al. 2022) have already been described in case studies and are not explained here again. The following will focus on a few cases that have not yet been published or that can be examined in more detail. The final evaluation of the countless single finds is not yet complete and will certainly yield further surprises and evidence of correlations in future research.

5.8.1 Case study I: HAM5 – *Danuvius guggenmosi* – the great apes from HAM5

The newly described great ape species *Danuvius guggenmosi* is likely one of the most renowned discoveries from the Hammerschmiede fossil site. The corresponding specimens were not found as an articulated skeleton, but represent reconstructed skeletal strewnfields over a distance of 30 metres length (Fig. 5.8.1.1).

For the taphonomic investigations of the Hammerschmiede sediments, a male *Danuvius* individual was one of the first and most important proofs that single specimens from the polyspecific bone accumulations of the fluvial Hammerschmiede deposits can be combined and reconstructed to individuals even when dispersed over such large distances. The arguments for assigning some of these bones to a single individual are based on correspondences in age- and size-specific characteristics but also on pathological consensus. Furthermore, the geospatial distribution of the bones and their arrangement within the suspected strewnfield of the reconstructed stream course of the HAM5 rivulet plays a crucial role in this assignment. Due to the specific density and surface characteristics of each skeletal element, their transport and depositional behaviour can vary significantly (Voorhies 1969, Behrensmeyer 1975, Hanson 1980, Behrensmeyer 1988). In the case of the male *Danuvius*, one of the crucial pieces of evidence that the bones belonged to one and the same individual is that the upper and lower jaws occlude perfectly. Since dental wear indicates a mature individual, the likelihood of this observation being a coincidence is reduced.

Furthermore, within the reconstructed strewnfield no duplicate skeletal elements were found. As with recent great apes, the hominid finds from Hammerschmiede show a similar size difference, with the present case corresponding to a male, while other finds are attributed to juveniles and females. Finally, the arrangement of the bones within the rivulet deposits of HAM5 corresponds fairly closely to the classical sorting by different bone properties (Voorhies 1969, Behrensmeyer 1975), which begins with the most dense elements such as teeth and jaws, through long bones, and finally (if still present at all) vertebrae and ribs (there are no ribs present in the case at hand) (Fig. 5.8.1.1). The vertebrae specimens, which are particularly important for the functional morphological interpretation of this new great ape species, indicate that the origin of the carcass and the transport distance cannot be too wide and probably lie in the area of the tooth finds, since vertebrae tend to be easily transported very far away due to their low density and voluminous appearance. This even allows flotation in the water body above certain flow velocities (Voorhies 1969, Behrensmeyer 1975, Hanson 1980).

5.8.2 Case study II: HAM5 – lateral overbank extension

The previously described bony strewnfield of a male *Danuvius guggenmosi* originates from the deeper main channel of the HAM5. It was not until 2018, passing over a slight ridge of elevation in the channel base, which was originally thought to be the channel margin, that a connected side channel was discovered in parallel to the west of this main channel. This side channel resembles a lateral overbank extension of the HAM5 and reaches only about half the depth of the main channel itself and probably correlates only with the third and final sedimentation phase to the main channel. In this area, it was possible to excavate fully to the actual western slope and edge of the

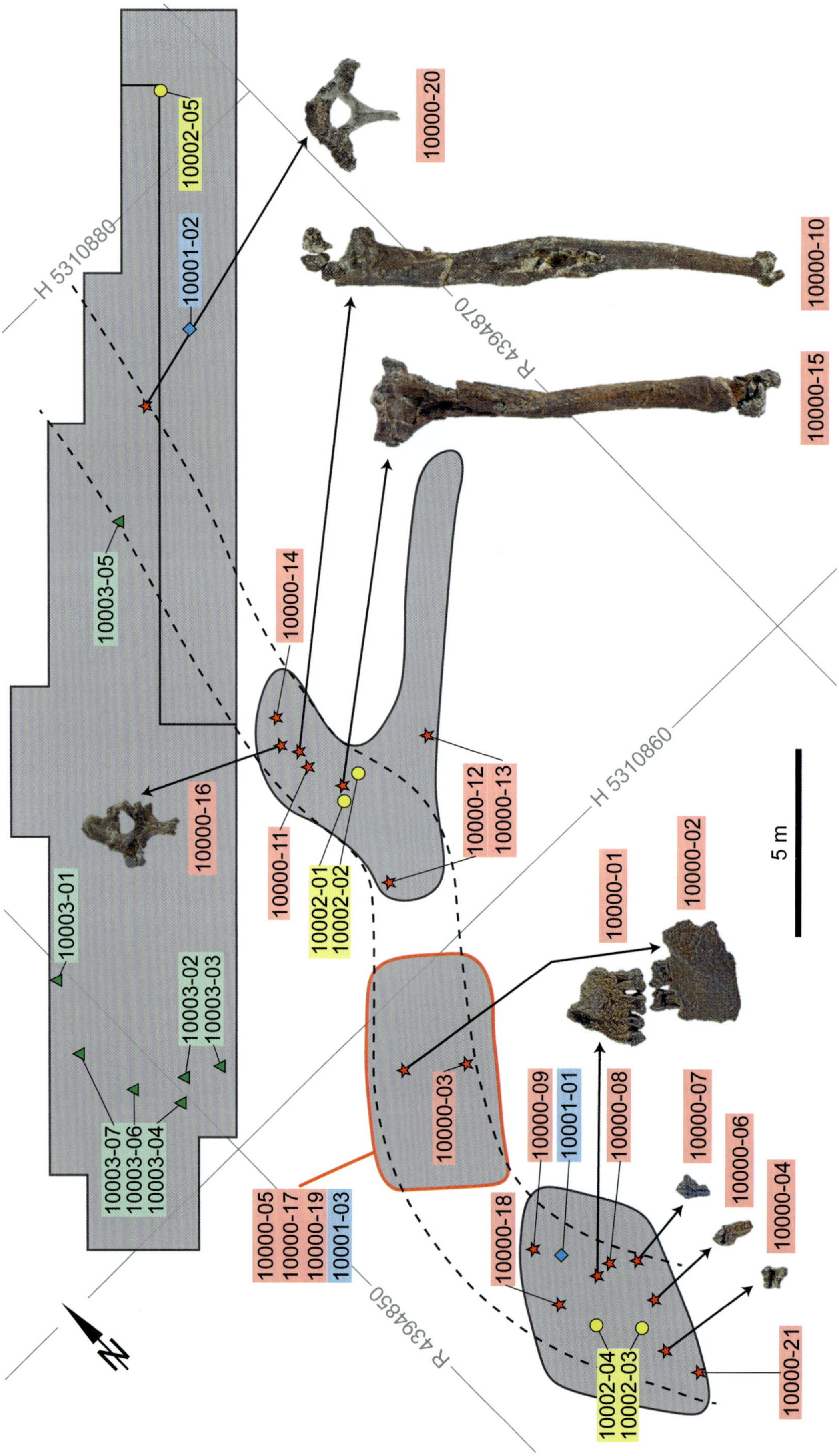

10000-20
10000-10
10000-15
10002-05
10001-02
H 5310880
R 4394870
10003-05
10000-14
10000-16
10000-11
10002-01
10002-02
10000-12
10000-13
H 5310860
5 m
10003-01
10003-02
10003-03
10003-07
10003-06
10003-04
10000-01
10000-02
10000-03
10000-09
10001-01
10000-08
10000-07
10000-06
10000-04
10000-05
10000-17
10000-19
10001-03
10000-18
R 4394850
N
10002-04
10002-03
10000-21

◄ **Figure 5.8.1.1.** Excavation plan with localized *Danuvius guggenmosi* specimens at the early Late Miocene Hammerschmiede deposits. Excavated areas of the local stratigraphic level HAM5 coloured in grey (excavation years 2015–2018). Intermediate regions represent material loss due to clay mining. Dashed lines indicate the supposed thalweg course of the palaeochannel with supposed palaeo-flow direction to the north. The different colours and symbols indicate assumed different *Danuvius* individuals: Supposed adult male individual (Holotype; GPIT/MA/10000; red stars); Female 1 (Paratype; GPIT/MA/10001; blue diamonds); Juvenile individual (Paratype; GPIT/MA/10002; yellow circles); Female 2 (Paratype, GPIT/MA/10003, green triangles). The area outlined in red shows the sediments removed and stored separately, which were later processed and screen-washed. To avoid mining-induced losses of specimens an imprecise localization of finds was accepted. Selected objects from the holotype and the most important strewnfield of the *Danuvius* finds are depicted according to their point of discovery. Coordinates correspond to Gauss-Krüger Zone 4 grid. Grid spacing equals 20 metres. Figure supplemented and revised based on Böhme et al. (2019a): Extended Data Fig. 1.

HAM5 channel from 2018 to 2022 and to document and measure all finds. In contrast to the main channel, this secondary section shows much less heterogeneity within the channel fill and fossil content. There were hardly any fin spines of catfish found and also turtle findings were much rarer, which characterise the classic enrichment layer of the HAM5 main channel. It is tentatively suggested that this side area does not represent the permanently flooded channel, but rather a lateral overbank extension. Likewise, the vertebrate fill shows what appears to be a much less reworked fossil content that can be attributed to a few single individuals of different mammalian species, which will be discussed in more detail below. The channel bottom in the mentioned area is particularly heterogeneous and characterised by small depressions and minor elevated regions. Some of these depressions in the fossil river bottom seem to have acted as fossil traps and have accumulated particularly many bones.

In the present case study of the western HAM5 channel area, several overlapping presumed skeletal strewnfields are evaluated in parallel. The selected cases represent the most prominent and well-preserved bone accumulations, each attributed to a single individual. A more detailed evaluation will certainly reveal further skeletal strewnfields in this area. In the following, the dispersed skeletons of a juvenile *Deinotherium*, an antelope, a large beaver, a large feliform, and two tragulids will be examined in greater detail, exemplifying the remarkable spatial proximity and overlap of individual skeletons in this area.

One of the most conspicuous skeletal strewnfields consists of 24 elements, all belonging to a juvenile (deciduous dentition) *Deinotherium levius*, which were found over a length of eight metres in the western part of the western overbank extension of the HAM5 channel (Fig. 5.8.2.1). Konidaris et al. (2023) described and mapped this find taxonomically and discussed the corresponding taphonomy in more detail. The find contains the mandible, several deciduous teeth, skull fragments, two long bones, the pubis, and 14 ribs. A large number of ribs, in particular, were excavated from the basal area of a deeper longitudinal erosional channel-base depression, which may represent a bone trap where the fossils accumulated in a flow shadow (Konidaris et al. 2023). The rarity of proboscidean bones in the Hammerschmiede deposits, the integrity in taxonomy and life age, the taphonomic distribution of finds, and the absence of duplicate elements in this case support the assignment of these specimens to one and the same individual (Konidaris et al. 2023). The arrangement of the finds with the objects of higher density, teeth and mandibles in the southernmost direction and other bones further north, supports and confirms the flow direction from the south already assumed in other studies (e. g., Böhme et al. 2019a).

In the same westernmost HAM5 area lies a somewhat more difficult to interpret distribution of bones. Over a length of (so far) 9.5 metres a strewnfield of bones extends, which belong to two adult tragulids (some duplicate elements) of the species *Dorcatherium naui* (Fig. 5.8.2.1). In the southern range, the high density of finds, including a partially articulated hind leg and skull, most likely represents a single individual. However, since at the northernmost excavated region another *Dorcatherium* skull was found, representing nearly the same dental wear and life age, including both mandibles possibly indicates the beginning of a second strewnfield of a *Dorcatherium* individual. The transitional area of the two skeleton distributions in particular cannot be clearly delimited, and an overlap of both strewnfields seems certain. At the moment, a more precise separation of the two strewnfields is not possible, and the find complex remains unresolved as a double-strewnfield. Beyond that, however, there is nothing to suggest that there is another adult *Dorcatherium* in this area. Even if a final separation is currently not possible, the distribution pattern of the *Dorcatherim* double-strewnfield fits very well with the general observations within the lateral overbank extension of HAM5. The finds, like those of the other strewnfields from this area, are distributed in the assumed flow direction from a southern direction. Bones of higher density such as long bones and cranial elements tend to remain in more southern spots, while lighter parts are transported to more northerly or distant regions (Voorhies 1969, Behrensmeyer 1975, Hanson 1980, Behrensmeyer 1988).

Likewise, in the western half of this lateral HAM5 area, there is a massive accumulation of bones, all of which can be taxonomically assigned to the large beaver species *Steneofiber depereti* (Fig. 5.8.2.1). There are

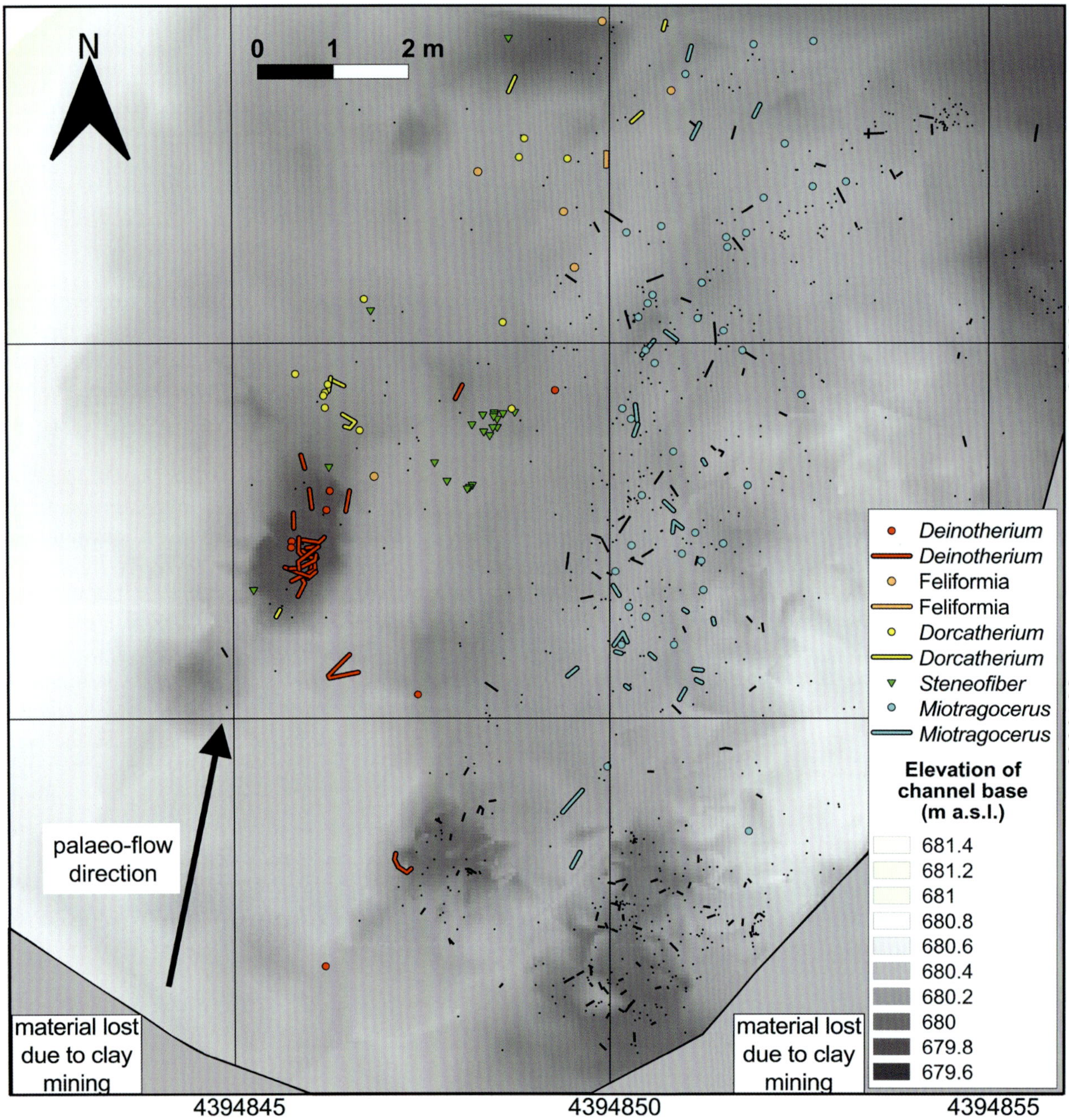

Figure 5.8.2.1. Distribution of finds in the western section of the excavation plan for the local stratigraphic level HAM5 at the early Late Miocene Hammerschmiede fossil site. The shown section comprises the lateral overbank extension of the western HAM5 channel excavated during the years 2018–2021. Black dots represent vertebrate fossil specimens, black lines denote elongated objects (long bones) and their orientation. Shaded areas have been lost to clay mining. The background colour indicates the differences in altitude of the excavated palaeochannel base: darker shades represent lower and lighter shades higher elevations, with a total elevation difference of 1.5 m (finds only in the lowermost 70 cm). Note how the specimen density partly corresponds to the uneven channel base, which probably acted as bone traps in pool-like structures. Five skeletal strewnfields (including one double strewnfield) of potential single individuals are shown in coloured symbols: a juvenile *Deinotherium levius* (24 elements; red circles and lines); a larger feliform carnivore (seven elements; orange circles and lines); presumably two adult tragulids (*Dorcatherium naui*) (double strewnfield; 20 elements; yellow circles and lines); a large beaver (*Steneofiber depereti*) (29 elements; green triangles) and a boselaphin antelope (*Miotragocerus monacensis*) (64 elements; light blue circles and lines). Black arrow indicates the assumed palaeo-flow direction of the palaeochannel and refers to long bone orientations, the channel morphology, and the extension of the presumed skeletal strewnfields. Coordinates correspond to Gauss-Krüger Zone 4 grid in metres. Grid spacing equals 5 metres.

25 bones from the hind limbs and from the forelimbs, which, including a few vertebrae, suggest an original carcass position of one adult beaver on an area measuring only a little more than one square metre. At the westernmost edge of this lateral HAM5 overbank extension, four more beaver bones were found, which can also be assigned to this individual with high probability due to their assumed life age (adult), the absence of duplicate elements, and the same bone size and proportions (e. g., left and right astragalus). The arrangement

of all 29 specimens follows a similar longitudinal pattern (south–north), as already observed by the previously described strewnfields. A special feature of this find is that there is not a single rib find in the area, and cranial elements that could be attributed to this individual are also missing. This and also the position of some bones further west could indicate that the initial carcass position was further south, outside the excavation area, as the cranial elements in particular should not be transported so easily due to their higher density (Voorhies 1969, Behrensmeyer 1975, Hanson 1980, Behrensmeyer 1988).

Seven isolated bones of a large adult carnivore were found dispersed very widely across the western half of the lateral overbank extension of HAM5, which so far can taxonomically be assigned to a feliform carnivore (Fig. 5.8.2.1). Due to the rarity of postcranial elements of large carnivores, the appropriate size and proportion, taxonomic classification, and the lack of duplicate elements, it can be assumed that this is the strewnfield of only one individual. The seven elements were found isolated and widely scattered over a distance of 12 metres (northern–southern extent) and mainly comprise limb elements. The isolated location of the specimens and the wide dispersion indicate a somewhat longer transportation route (Voorhies 1969, Behrensmeyer 1975, Hanson 1980, Behrensmeyer 1988). No direction of origin can be deduced from this find, but the strewnfield orientation in a south–north direction is consistent with the other strewnfields already observed at this area.

One of the probably most extensive bone distributions is located in the eastern part of the lateral overbank extension of HAM5. 64 objects have already been found over a distance of 21 metres, which probably represent a single individual of an adult antelope of the species *Miotragocerus monacensis* (Fig. 5.8.2.1). The corresponding skull of this individual was already described by Hartung et al. (2020). The assignment of the specimens is performed on the basis of taxonomic classification, bone sizes, and proportions, and also no bone duplication (based on bone preservation and life age) is present. The bone preservation shows a pronounced tendency toward severe compression and flattening of many elements, likely due to their position within a highly compressive, thin, and ductile clay layer (see Fig. 5.6.2.1 b). Particularly dense bones (long bones) and the cranial elements (skull fragments and lower jaws) are distributed within the southern third, while smaller bones like vertebrae or ribs are scattered over a greater distance to the north. According to this arrangement, which is typical for skeletal strewnfields, a transport direction to the north and an initial deposition of the carcasses at a southern position seems very certain (Voorhies 1969, Behrensmeyer 1975, Hanson 1980, Behrensmeyer 1988).

5.8.3 Case study III: HAM4 – *Propotamochoerus palaeochoerus*

In the HAM4 excavation area during the 2017 and 2018 field seasons, a remarkably well-preserved skull of a juvenile female suid, identified as *Propotamochoerus palaeochoerus*, was discovered (Fig. 5.8.3.1). Within a one-metre radius of the cranium, a mandible and various isolated teeth were recovered, which correspond precisely to the missing tooth positions of the skull and fit into the alveoli (Fig. 5.8.3.2).

Based on consistent taxonomic identification, tooth wear patterns, the stage of tooth eruption, and the absence of duplicate elements, all of these remains are confidently assigned to the same individual. The cranial elements were positioned at the top of the uppermost in a series of inclined layers within the lower point bar. These layers exhibit fining-upward sequences of large-scale trough-cross-bedded fine sands, which originate from basal, coarse-grained deposits composed of reworked pedogenic carbonates and mud clasts. The highest skeletal elements were found approximately 40 cm above the channel base, while the deepest lie directly on the base itself (Fig. 5.8.3.2). The entire assemblage was overlain by trough-cross-bedded fine sand, characterised by reddish streaks (schlieren) of garnet and other heavy mineral accumulations.

Since these skeletal elements are distributed across a substantial vertical range, their preservation holds particular taphonomic significance. Further excavations in the southern sector of the site yielded additional postcranial elements, which are also attributed to the same individual based on multiple lines of evidence. Just a few centimetres south of the skull discovery, an arrangement of four ribs in consecutive anatomical positions on the left body side was found. Over the remaining excavation area, another 14 scattered ribs were found, all of which match the juvenile suid in size, ontogenetic stage, and preservation state.

Morphologically, all ribs from the anterior and medial thoracic regions share a small additional process below the capitulum, a feature that may be unique to suids or even exclusive to *Propotamochoerus palaeochoerus*. Notably, no other mammal from the Hammerschmiede deposits has been identified with this characteristic.

At some distance to the south, articulated left ulna and radius (GPIT/MA/16642) as well as a right ulna (GPIT/MA/16641) were recovered (Fig. 5.8.3.3). All of these elements correspond to a small, juvenile suid, matching the young female skull in terms of size, ontogenetic stage (absence of epiphyses and unfused epiphyseal joints), and preservation. Further south, over 13 metres from the cranial remains, a scapula (GPIT/MA/16625) and two consecutive ribs in anatomical position (GPIT/MA/16635) mark the most

Figure 5.8.3.1. Skull of the juvenile suid *Propotamochoerus palaeochoerus* from the local stratigraphic level HAM4 at the early Late Miocene Hammerschmiede fossil site. The excellently preserved skull was one of the first finds from a skeletal strewnfield, which has since been reconstructed over a length of around 15 metres in the HAM4 excavation area (2017–2018). The lower jaw (GPIT/MA/12500-2) and skull (GPIT/MA/12500-1) were digitally reassembled, as the original pieces cannot be put together directly due to slight deformation. Scale bar equals 5 cm.

distant found locations. Unfortunately, due to ongoing clay mining, the area beyond this point was lost before excavations could continue, leaving open the possibility that additional skeletal elements once extended farther south.

Across the entire assemblage, no taxonomic duplication is observed – even among the 18 recovered ribs, each is unique, with no overlapping elements in terms of size, preservation, or ontogenetic stage. With an overall dispersal distance of 15 metres, this assemblage represents one of the most extensive strewnfields documented at the HAM4 level. Nowhere else in HAM4 has suid material of this exceptional preservation quality been recovered. This is remarkable because suid fossils at HAM4 are typically found as isolated specimens rather than in such a dense and concentrated depositional context. The exceptional state of preservation of all specimens assigned to this juvenile suid suggests that the carcass remained largely intact until shortly before final burial.

The strewnfield is located along the eastern outer edge of a presumed leftward meander in the HAM4 river course (Fig. 5.8.3.3). The relatively extensive strewnfield contains fossils found at higher elevations – up to 60 cm above the base – as well as material resting directly on the channel floor, indicating that the carcass is distributed across both the lower point bar and the channel lag.

Based on the relative positions of skeletal elements and referring to taphonomic studies by Voorhies (1969), Behrensmeyer (1975, 1988), and Hanson (1980), the close proximity of dense, heavy cranial remains and the wider dispersal of lighter bones (e. g., ribs) initially suggests a transport and deposition direction from north to south. However, as exemplified above, multiple independent lines of evidence confirm that the HAM4 river generally flowed from south to north.

A plausible explanation for this apparent contradiction is that: (1) The intact skull, including the mandible and dentition, was transported as a whole down towards the channel lag of the river's thalweg in the north, where it subsequently disarticulated. (2) The ribs, which would typically be more easily transported, may have remained anchored in place on the point bar due to its finer-grained, muddier substrate. The cohesive properties of these silt-rich and soft sediments may have caused the bones to adhere to the substrate, preventing further transport. (3) The absence of vertebrae within the strewnfield supports the hypothesis that those bulkier, more voluminous elements were more susceptible to transport and were carried away, while flatter bones such as ribs remained trapped in the muddy substrate. This effect is likely enhanced by near-bed low-velocity currents, especially over smooth substrates, where a sublayer can decouple turbulent flow and leave shallow objects undisturbed (e. g., Behrensmeyer 1975).

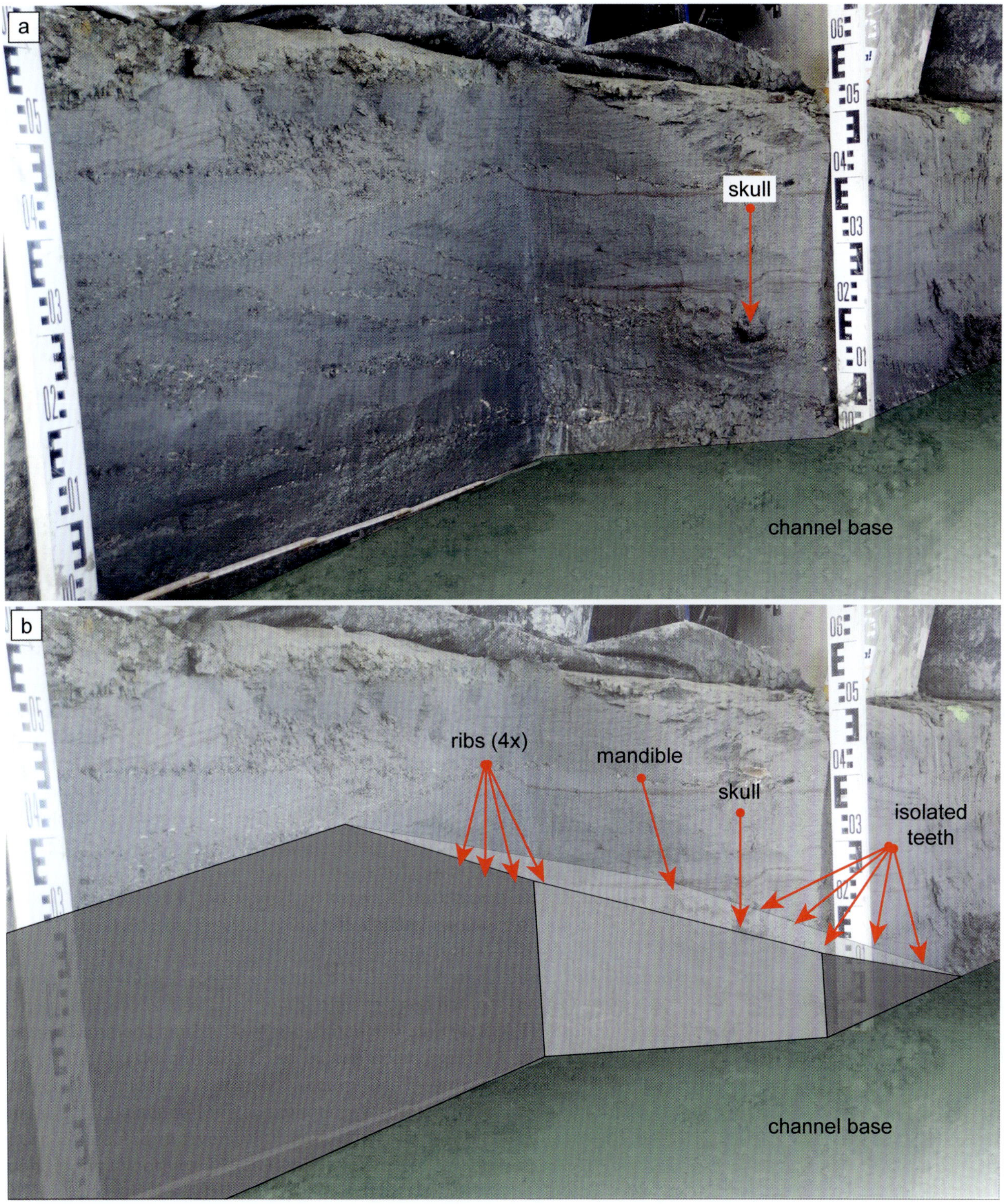

Figure 5.8.3.2. Find situation of the skull and associated finds of the juvenile suid *Propotamochoerus palaeochoerus* from the local stratigraphic level HAM4 at the early Late Miocene Hammerschmiede fossil site (September 2017). Cranial specimens and four ribs were located on top of the last in a series of inclined layers of the lower point bar, characterised by fining-upward sequences of large-scale trough cross-bedded fine sands initiated by coarse-grained reworking layers of pedogenic carbonates and mud clasts. The find surface inclines from a level of approx. 40 cm down to the channel base. Overlaying trough cross-bedded fine sand including reddish streaks (schlieren) of garnet indicates that the suid was deposited in a quick and final burial. (a) View of the profile face with labelled skull looking to NW. (b) Same image with overlays of the inclined find stratum and the approximate arrangement of the specimens. Scale shown with folding metre rule (20 cm) and metric scale bars (cm).

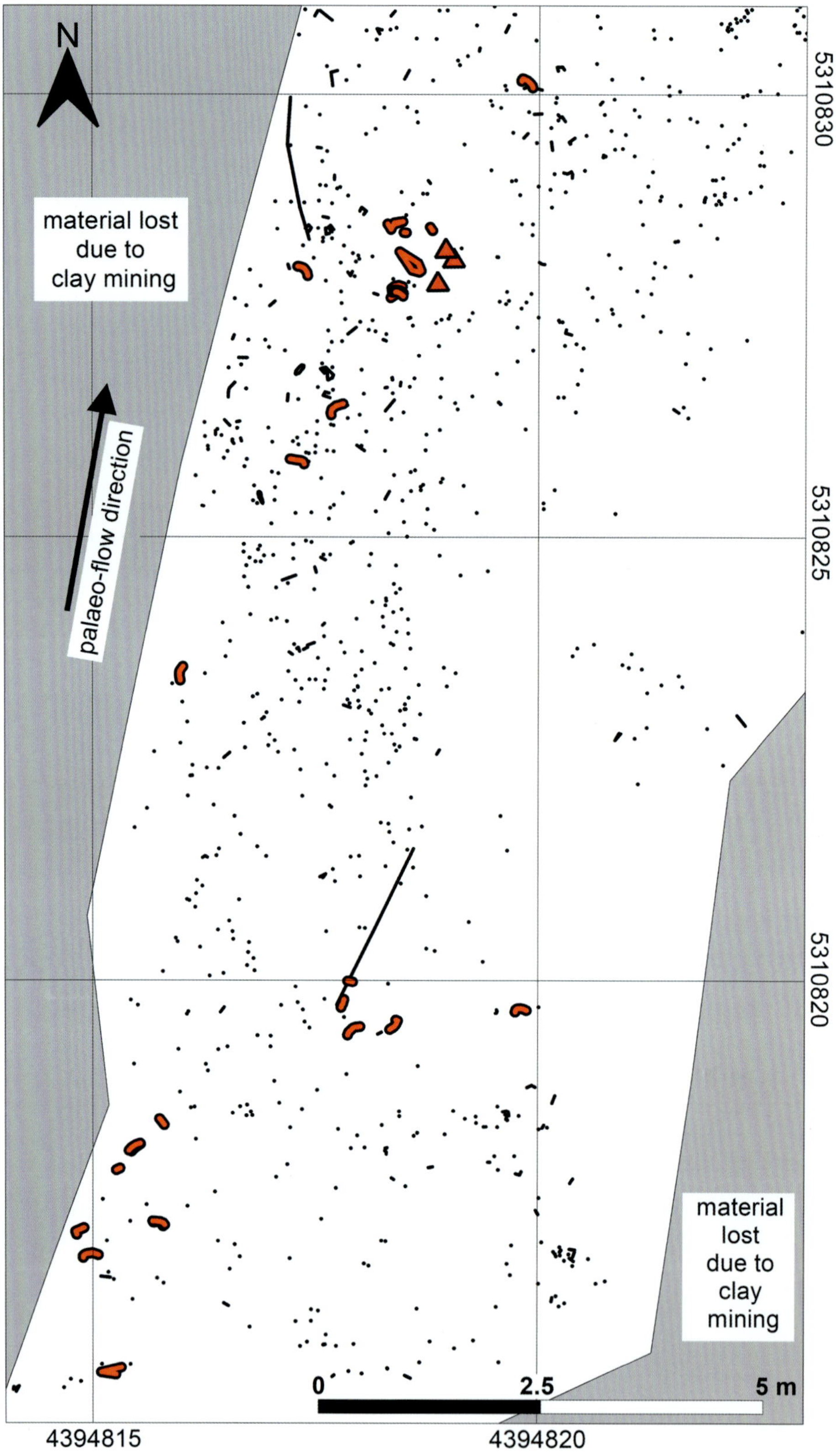

Figure 5.8.3.3. Section of the excavation plan at the local stratigraphic level HAM4 at the early Late Miocene Hammerschmiede fossil site. The shown area comprises the lateral eastern margin of the HAM4 channel excavated during the years 2017–2018. Black dots represent vertebrate fossil specimens, black lines denote elongated objects (the longest of these represent wood finds) and their orientation. Shaded areas have been lost due to clay mining. Excavated suid remains (30 specimens), most probably belonging to the same juvenile, female *Propotamochoerus palaeochoerus* individual, are highlighted with red triangles, lines, and outlines (skull). Associated bones (limb bones and mostly ribs) are arranged over a distance of 16 metres parallel to the reconstructed flow direction (black arrow). An enrichment of cranial elements (skull with mandible and associated isolated teeth), as well as partially articulated/arranged ribs is found in the northern strewnfield area. Coordinates correspond to Gauss-Krüger Zone 4 grid in metres. Grid spacing equals 5 metres.

5.8.4 Case study IV: HAM4 – *Dorcatherium naui*

The tragulid *Dorcatherium naui* is among the most abundant large mammal remains found in the Hammerschmiede deposits. Since finds of *Dorcatherium* are scattered almost everywhere, it is particularly difficult to reconstruct single individuals by strewnfields. In the southernmost excavation area (2019) within the very small proximity of a little more than one square metre at least 53 bones have been unearthed that are all anatomically assigned to the tragulid species *Dorcatherium naui* (Fig. 5.8.4.1).

The western section consists of various bones of the left and right hind limbs (without pelvic bones), and the eastern section yields 20 ribs (eleven right, nine left) from an extremely small spot (Fig. 5.8.4.2). In the entire area one thoracic vertebra and four cervical vertebrae (including axis but no atlas) were found. Between the closely adjacent find areas of the hind limbs and the ribs an excellent preserved skull of an adult female *Dorcatherium naui* was recovered (Hartung & Böhme 2022). Furthermore, a hemimandible was excavated in very close proximity to the main accumulations, which matches the dental wear and state of the cranium and occludes perfectly with the maxillary tooth row.

The monospecific accumulation of bones of *Dorcatherium naui* and the low density of finds in the vicinity of this enrichment spot, the anatomically correct arrangement (no perfect articulation) of many bones and skeletal regions, the perfect fit of pedal joints and the absence of duplicate skeletal elements leads to the conclusion that all of these bones belong to a single individual.

The present case is the first observation of a special taphonomic phenomenon in the Hammerschmiede deposits. The classical theory assumes that sedimentation or transport of bones is controlled by their physical properties, and especially by the density and shape of the bones (Voorhies 1969, Behrensmeyer 1975, Hanson 1980, Behrensmeyer 1988). According to this model, ribs and vertebrae are the "lightest" elements, which are transported the fastest and farthest away from the initial point a carcass enters the fluvial system. In this case study, ribs from both sides of the thorax appear to have remained at their initial location, along with the skull and some limb bones, while the thoracic vertebrae are largely absent, suggesting they were transported elsewhere. This biostratinomic observation is somehow similar to the situation of the juvenile *Propotamochoerus* described above, and we favour a similar explanation. A decisively different parameter to the classical observation models is that the Ham-

Figure 5.8.4.1. Reconstruction of the skeleton of *Dorcatherium naui* (Artiodactyla, Tragulidae) from the early Late Miocene Hammerschmiede fossil site and the local stratigraphic level HAM4. The skeleton was digitally assembled from 53 bones, which presumably belong to one individual. Bones belonging to the skeleton find, which are only present on one body side, were mirrored. Supplemented elements in white. Scapula, for *Dorcatherium* from Hammerschmiede unknown at the time of reconstruction, is supplemented from living artiodactyles (black). The front legs do not belong to this individual but were added from finds from the Hammerschmiede. Scale bar equals approximately 10 cm. Reconstruction created by C. Kyriakouli.

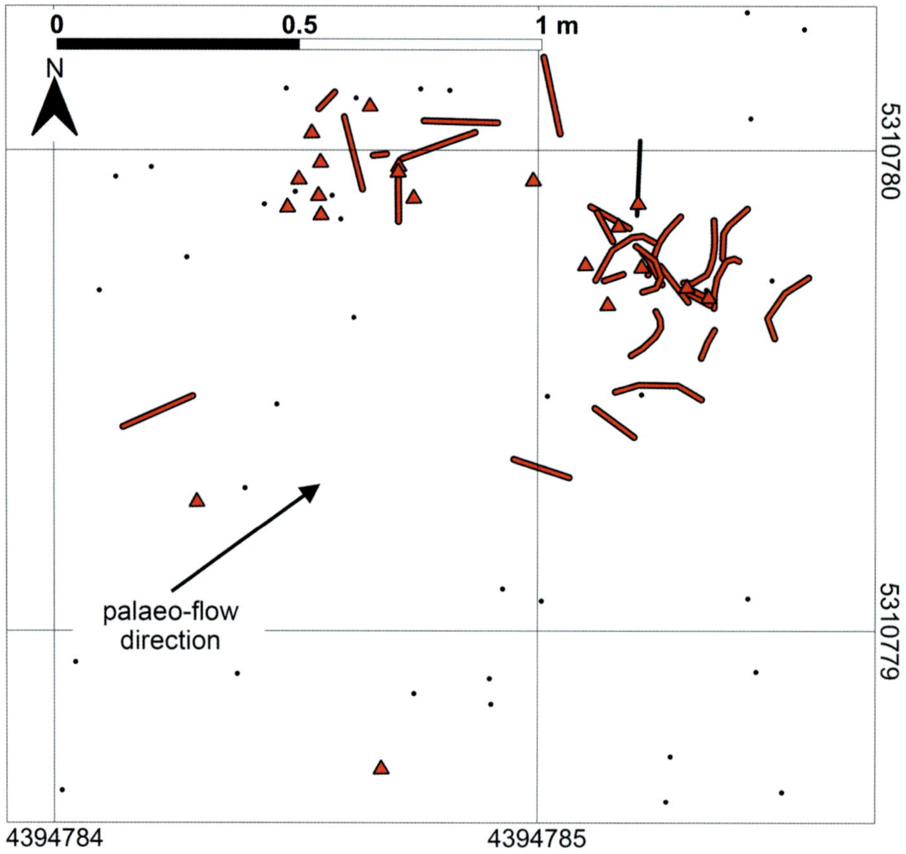

Figure 5.8.4.2. Section of the excavation plan of the local stratigraphic level HAM4 at the early Late Miocene Hammerschmiede fossil site. The shown area comprises the part of the southernmost excavation area near to the deepest central channel area of the HAM4 channel excavated in 2019. Black dots represent vertebrate fossil specimens, black line denotes an elongated bone and orientation. Excavated tragulid remains (53 specimens), most probably belonging to the same adult, female Dor*catherium naui* individual, are highlighted with red triangles and lines. Associated bones (limb bones and mostly ribs) are arranged over a very small area (approximately one square meter) and long bone orientations correspond to the reconstructed flow direction (black arrow) for this area. Coordinates correspond to Gauss-Krüger Zone 4 grid in metres. Grid spacing equals one metre.

merschmiede deposits comprise fine-grained and more soft sediments (clay, silt, and fine sand). These muddy sediments could simply have adhesively attached the bones to the ground. The anatomical arrangement of the bones indicates that some decomposition must have taken place, separating the bones from each other but perhaps also partially holding them in tendon union. Remarkably, all finds where a reconstruction to a strewnfield is possible do not show any signs of feeding marks, so it can be assumed that the place where the carcasses decayed was not accessible to predators such as hyaenas, and this was possibly within the river. Furthermore, the carcass decomposed without great rearrangement processes acting. The carcass disintegrated and collapsed, with all the bones settling on the lower point bar. During a subsequent event, an increase in flow velocity and a rise in river level further influenced their distribution. The basal-most currents are very low due to friction, and particularly in smooth riverbeds a low-velocity sublayer can form which separates the turbulent flow and thus leaves mainly shallow objects on the bottom untouched (e.g., Behrensmeyer 1975). Bones that are somewhat more voluminous and protrude further into the flowing current (e.g., eddies) have considerably more surface area to be attacked by the eddies and to be detached from the muddy bottom. In contrast to thoracic vertebrae, which have particularly long neural spines, ribs are very flat and tend to remain adhesively attached to the soft bottom sediments, stay in place, and eventually become quickly buried in sediment and thus further stabilised.

As a result of this observation, adhesion is supposedly one of the primary physical parameters that influence the retention or detachment of bones from a river bottom consisting of fine-grained and thus muddy sediments as it is the case of the Hammerschmiede deposits. Accordingly, the fluid dynamic behaviour of a bone is superimposed by the bone-substrate interaction and determines whether this enables detachment or whether the bone remains in place and becomes deposited. In this context, the shape and density of the surface to be attacked by the acting forces and their relationship to the adhesive suction of a bone surface to the underlying channel sediments also play a decisive role.

In this case study, the Hammerschmiede site contributes significantly to the understanding of transport and retention of bones in rivers with fine and rather muddy sediments. Understanding these special channel bottom properties seems to be essential for understanding the dispersion and distribution of the skeletal strewnfields at the Hammerschmiede fossil sites and beyond that.

5.8.5 Case study V: HAM4 – *Miotragocerus monacensis*

The boselaphin bovid *Miotragocerus monacensis* is, alongside tragulids, one of the most common large mammals found in the HAM4 channel deposits. In general, isolated bones and teeth of *Miotragocerus* are interspersed throughout the entire channel fill. Due to the presumed high number of intermixed individuals, assigning isolated specimens to individual carcasses is challenging. However, two notable find distributions stand out due to their exceptionally high specimen density and spatial context, making a connection to single individuals highly probable.

Excavations in 2021 and 2022 in the central HAM4 channel revealed two spatially confined zones with particularly high concentrations of limb bones, all of which belong to *Miotragocerus monacensis*. This assemblage includes at least 34 documented bones, excavated from a northern (22 specimens) and a southern zone (12 specimens). These two concentrations are approximately 12 metres apart (Fig. 5.8.5.1).

The northern site covers an area of approximately 2 × 3 metres, elongated in a north–south direction. Within this zone, several non-articulated but partially anatomically arranged *Miotragocerus monacensis* bones were found, including a left anterior limb, both left and right hind limbs, the pelvis, the sacrum, and some tail and lumbar vertebrae (Fig. 5.8.5.2). These finds are embedded within the basal sediments of the HAM4 channel's lower bar units, positioned 10–50 cm above the channel base. Although not articulated, the correct anatomical alignment suggests that these remains likely belong to a single individual. This interpretation is further supported by the notably high concentration of long bones – a feature atypical of Hammerschmiede find layers – as well as the absence of duplicate skeletal elements. Additionally, the bones closely correspond in size, proportion, ontogenetic stage, and preservation state.

A second smaller find site to the south covers less than 0.5 square metres, where an anatomically correctly arranged (not articulated) right forelimb of a *Miotragocerus monacensis* was discovered (Fig. 5.8.5.3). The elevation of this find matches that of the northern locality, being positioned slightly above the channel base within the lower bar units. The completeness, anatomical arrangement, matching ontogenetic stage, and preservation quality confirm that it represents a single isolated limb from the same antelope species.

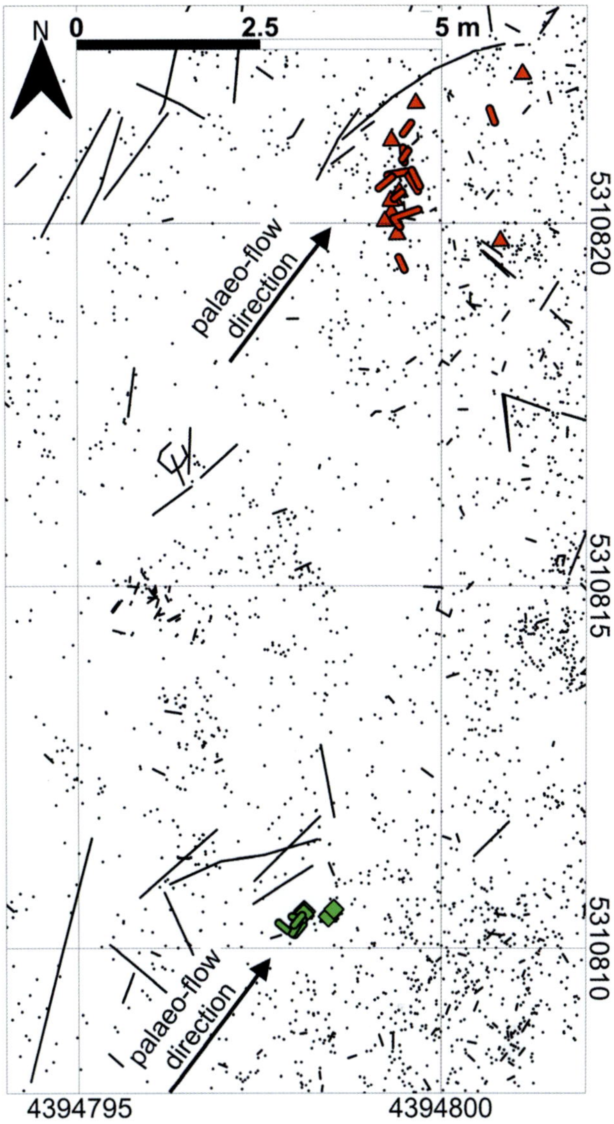

Figure 5.8.5.1. Section of the excavation plan of the local stratigraphic level HAM4 at the early Late Miocene Hammerschmiede fossil site. The shown area comprises a part of the central excavation area near the deepest channel area of the HAM4 excavated in 2021 and 2022. Black dots represent vertebrate fossil specimens, black lines denote elongated objects (the longest of these represent wood finds) and their orientation. Associated bones of *Miotragocerus monacensis* (34 specimens) are arranged within two isolated small-scale strewnfields approximately 12 metres apart from each other. Those spots comprise a very high bone density consisting of several elements belonging to both hind limbs and the left forelimb in the north (red triangles and lines) and a partially articulated and arranged right forelimb in the south (green diamonds and lines). Coordinates correspond to Gauss-Krüger Zone 4 grid in metres. Grid spacing equals 5 metres.

The localized nature of these high-density strewnfields suggests that each assemblage corresponds to individual *Miotragocerus* carcasses. In the broader context, both sites stand out due to their high specimen density, with a notable lack of intervening finds between the two spots.

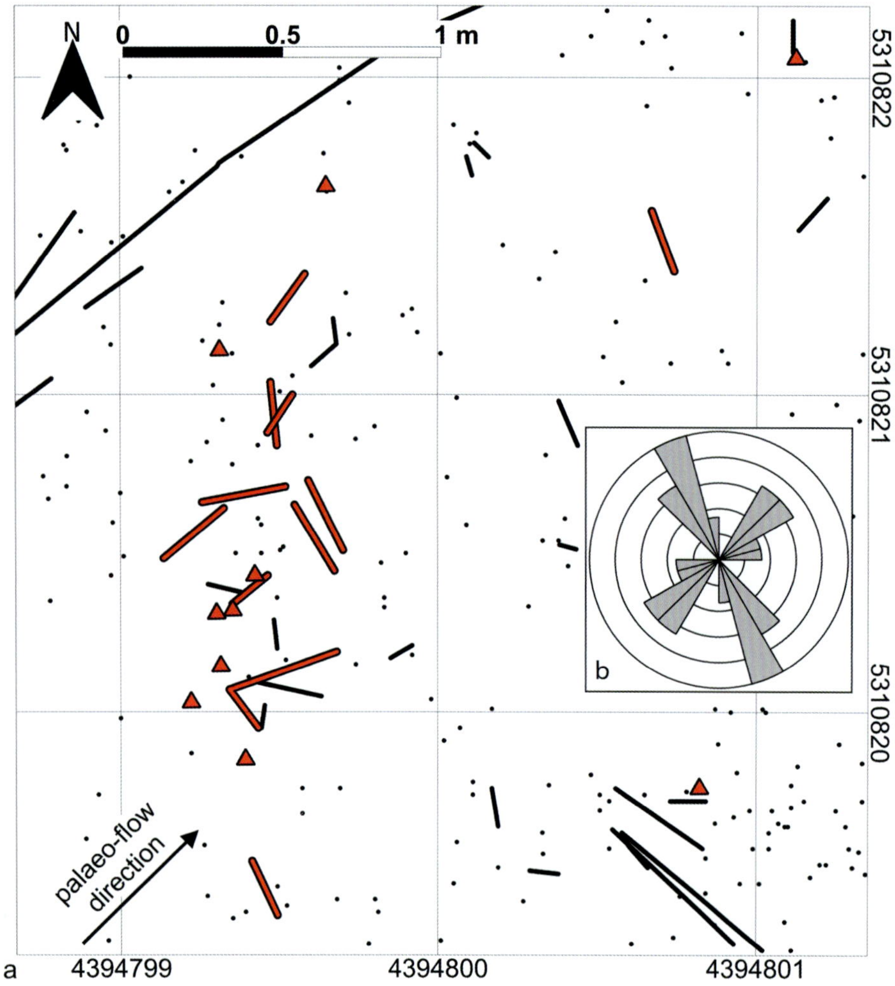

Figure 5.8.5.2. (a) Detailed view of the northern bovid strewnfield section of the excavation plan of the local stratigraphic level HAM4 at the early Late Miocene Hammerschmiede fossil site. The shown area comprises a part of the central excavation area near the deepest channel area of the HAM4 channel excavated in 2021 and 2022. Black dots represent vertebrate fossil specimens, black lines denote elongated objects and their orientation. Excavated bovid remains of the northern strewnfield section, most probably belonging to the same *Miotragocerus monacensis* individual, are highlighted with red triangles and lines. Associated bones (mostly limb bones) are arranged and oriented in correspondence to the reconstructed flow direction (black arrow) in this area. The shown accumulation consists of several elements belonging to both hind limbs in addition to the pelvic region and lumbar vertebrae and the left forelimb. (b) Line direction histogram of elongated bovid bones of the strewnfield indicates the typical bone alignment pattern in the longitudinal and transverse direction to the flow. Coordinates correspond to Gauss-Krüger Zone 4 grid in metres. Grid spacing equals one metre.

Although the two strewnfields are separated by 12 metres, they exhibit remarkable similarities in taxonomy, age, preservation state, articulation pattern, and complementarity. It is conceivable that the southern right leg corresponds to the northern left leg, placing the two find zones in a shared depositional context. However, this hypothesis requires further investigation.

A more detailed analysis of depositional patterns challenges the assumption that these remains belong to a single strewnfield. Observing the orientation of long objects (bones and wood) within the excavation area reveals a prevailing deposition direction of SW–NE (Fig. 5.8.5.1). In contrast, the two *Miotragocerus* assemblages are positioned in a roughly S–N alignment. Furthermore, the bone orientations within the find zones themselves do not align with a direct fluvial transport pathway (Figs 5.8.5.2, 5.8.5.3). Thus, deposition in a single transport event appears unlikely.

Furthermore, the high specimen densities within the plots imply a deposition shortly before the final burial. The state of bone preservation also suggests a rapid embedding. This rules out secondary reworking and lateral displacement after deposition.

A profile section along the presumed flow direction further contradicts the idea of a single strewnfield. The longitudinal HAM4 profile reveals slightly inclined sedimentary packages, interpreted as downstream-accreting meander point bars (Fig. 5.8.5.4 and section 3.2). Although the two strewnfields share a similar elevation (just above the channel base), the profile demonstrates that each set of remains is associated with the basal deposit of a separate point bar unit (Fig. 5.8.5.4). This clear stratigraphic separation indicates that the

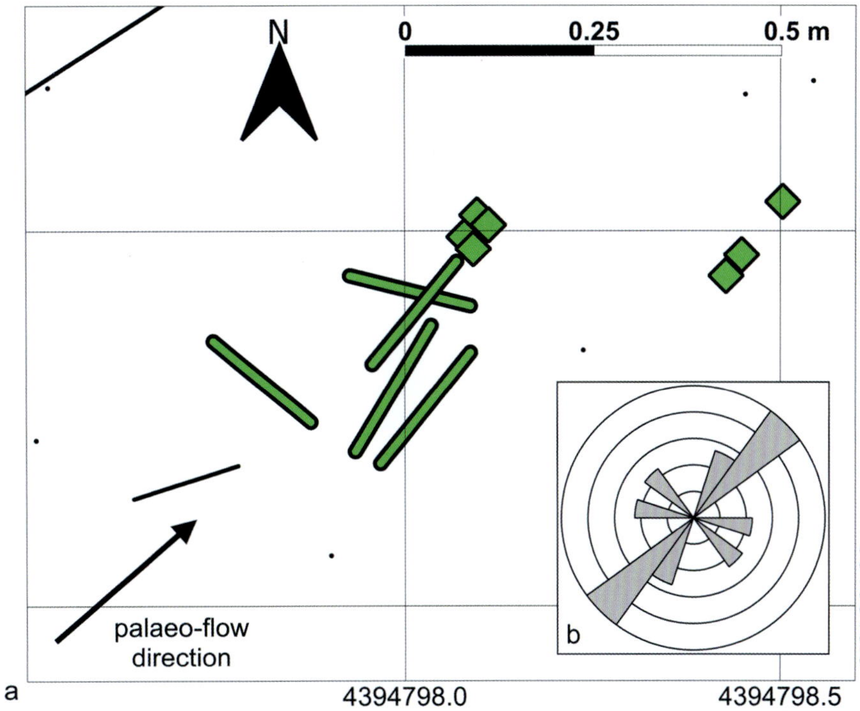

Figure 5.8.5.3. Detailed view of the southern bovid strewnfield section of the excavation plan of the local stratigraphic level HAM4 at the early Late Miocene Hammerschmiede fossil site. The shown area comprises a part of the central excavation area near the deepest channel area of the HAM4 channel excavated in 2021 and 2022. Black dots represent vertebrate fossil specimens, black lines denote elongated objects and their orientation. Excavated bovid remains of the southern strewnfield section, most probably belonging to the same *Miotragocerus monacensis* individual, are highlighted with green diamonds and lines. Associated bones belong to an isolated right forelimb and are arranged and oriented in accordance to the reconstructed flow direction (black arrow) at this area. Coordinates correspond to Gauss-Krüger Zone 4 grid in metres. Grid spacing equals 0.5 metres.

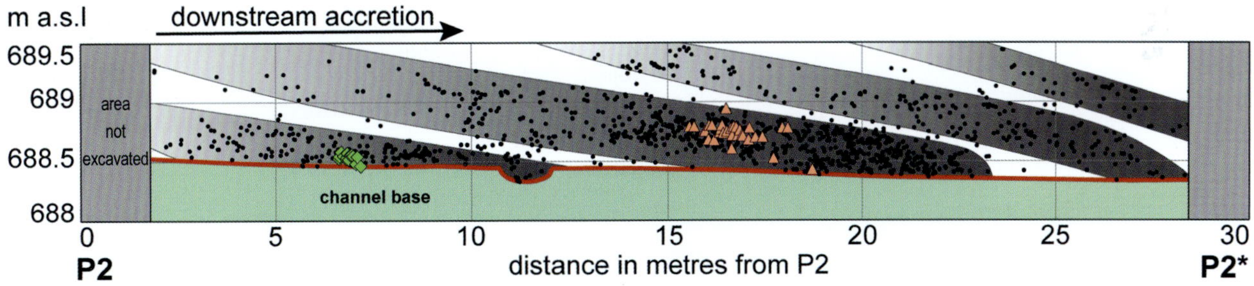

Figure 5.8.5.4. Accumulated longitudinal channel profile section including two bovid bone accumulations (green diamonds and red triangles) in the excavation area of the local stratigraphic level HAM4 at the early Late Miocene Hammerschmiede locality. The section is south-north oriented, looking to the north, and comprises data accumulated during excavations from 2019 to 2023 from a four metres wide transect along a median profile line from P2 (E4394793.83, N5310804.81) to P2* (E4394806.88, N5310832.25) (for profile location in the map see Fig. 3.2.3). Excavated specimens (back dots) are plotted according to their elevation above the HAM4 channel base (red line). Grey shaded overlays indicate the location of inclined packages (dip angle ranges from 2° in the south to 3° in the north), interpreted as downstream accretion of a meander point bar. The two strewnfields are clearly evident in two different point bar packages and thus also in temporal deposition distance to each other, which excludes a connection to the same individual. Easting coordinates correspond to Gauss-Krüger Zone 4 grid in metres and elevation in m a.s.l.

southern strewnfield (green diamonds, Fig. 5.8.5.4) represents an earlier deposition event, whereas the northern strewnfield (red triangles) corresponds to a later depositional phase.

This case exemplifies the complexity of depositional processes within the HAM4 system and highlights the importance of detailed stratigraphic analysis when evaluating skeletal assemblages.

Unlike other skeletal accumulations in Hammerschmiede, neither of the strewnfields contain assigned ribs nor any reliably attributable cranial elements. Given the excellent bone preservation, a fully intact cranium would be expected if the carcass had undergone minimal disturbance. However, no *Miotragocerus* skull has yet been found within these assemblages.

In other Hammerschmiede strewnfields, skulls and

ribs are often found at the presumed initial deposition site. Therefore, it is possible that the primary deposition site of these individuals has not yet been excavated. If this were the case, the currently known strewnfields would represent only the distal portions of a more extensive skeletal distribution, with the initial accumulation located farther south in unexcavated areas.

The presence of non-articulated but anatomically arranged bones suggests that the carcasses had reached a ligamentous stage of decomposition at the time of transport. Tendons and ligaments likely remained intact, preventing full skeletal disarticulation, even during prolonged fluvial transport. This cohesion ultimately ensures that the entire bone arrangement can be transported away from the original carcass location, and the bones remain together. The absence of carnivore bite marks further suggests that the remains were not scavenged, possibly because they were not easily accessible to predators. Although fluvial transport likely played a role in deposition, the findings lack clear evidence of prolonged high-energy transport. The most plausible explanation is that the remains entered the river from a nearby terrestrial setting (potentially the upper bar), where initial desiccation preserved ligament connections. However, their eventual burial in channel deposits suggests significant fluctuations in HAM4 river discharge and current intensity.

5.8.6 Case study VI: HAM4 — reassembled fossil broken and scattered bones

One of the best ways to obtain information about flow direction and potential individual dispersal fields are fragments that prove affiliation over long distances by an exact break edge match. So far only a few of these cases are documented, but in most cases there are only a few metres distance between the components. The problem of this method is the ever-increasing amount of material at a greater comparative distance to find the counterpart to a fragmented object. There is one case of a mammalian fibula diaphysis represented by two similarly sized pieces which were found about 1.5 metres apart from each other within the HAM4 excavation area of 2019-2 (Fig. 5.8.6.1). The alignment of the two fragments indicates a transport direction in north–south orientation in the HAM4 channel.

A second case, and one of the most comprehensive examples of reassembled bone fragments found over a larger distance, is a proboscidean rib refit of three larger isolated pieces found within the southernmost excavation area of the HAM4 channel (HAM4 2019-1). In this case, due to the rarity of this animal group, the comparative material is manageable and led by chance to the result. This case proves via direct fracture fit the affiliation of three each approx. 15 cm long

Figure 5.8.6.1. Scattered found and reassembled fragments of a mammalian fibula diaphysis from the early Late Miocene Hammerschmiede deposits and the local stratigraphic level HAM4. Both pieces were found about 1.5 metres apart and fit together perfectly at the fracture seam (a–b). (a–b) represent different view angles of the refit of the objects GPIT/MA/17170 and GPIT/MA/17180. Scale bar equals 10 mm.

Figure 5.8.6.2. Scattered and reassembled large fragments of a proboscidean rib, recovered from the early Late Miocene Hammerschmiede deposits at the local stratigraphic level HAM4. The three specimens were found approximately 2 metres apart from each other but show perfect but slightly rolled and polished fracture seams (a-c). (a) GPIT/MA/16784; (b) GPIT/MA/16806; (c) GPIT/MA/16749. Scale bar equals 5 cm.

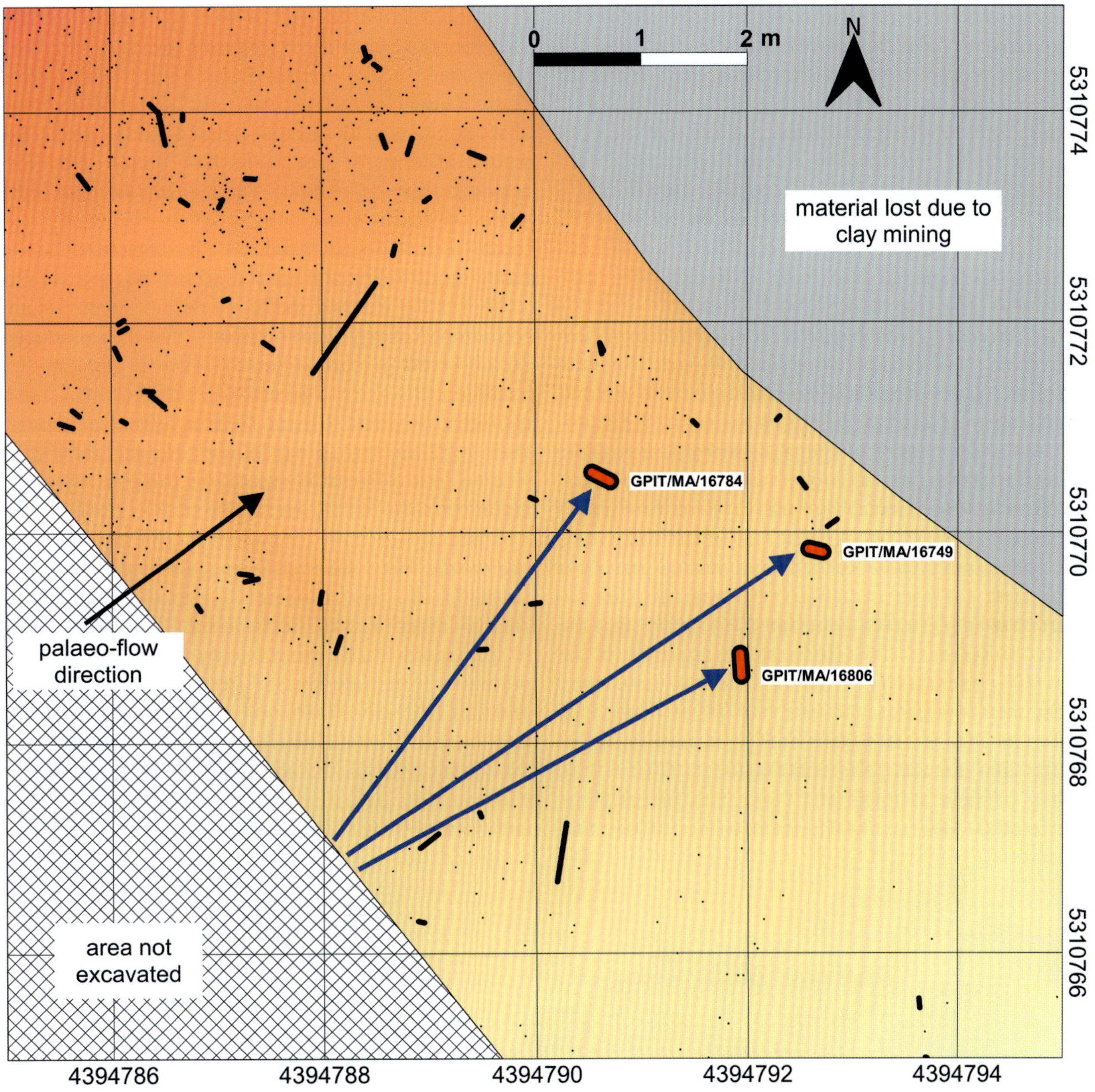

Figure 5.8.6.3. Detailed view of three scattered and reassembled large fragments of a proboscidean rib, found at the southernmost section of the excavation plan of the local stratigraphic level HAM4 at the early Late Miocene Hammerschmiede fossil site. The three specimens (GPIT/MA/16784, 16806 and 16749) were found approximately 2 metres apart from each other in lateral disposition and one metre north-south offset and show perfectly fitting but slightly rolled and polished fracture seams. A transport origin and a lateral dispersion are indicated from a most probably further southern initial point of fragmentation (blue arrows). Black dots represent other vertebrate fossil specimens, black lines denote elongated objects and their orientation (the longest of them are wood finds). Coordinates correspond to Gauss-Krüger Zone 4 grid in metres. Grid spacing equals 2 metres.

rib fragments of a large proboscidean, fitting together to a 58 cm long fragment (Fig. 5.8.6.2).

The individual parts were found over an area of several metres and were not arranged linearly. The rather triangular relation of the individual bones gives rise to the assumption that the fragmentation and the original starting point of the rib were further upstream and all three elements were distributed slightly independently of each other so that there is a laterally displaced position of each object (Fig. 5.8.6.3). Furthermore, both ends of the reassembled rib show breaks that likely occurred prior to or during final burial. This means that there may have been other pieces in the nearer and further surroundings, and that they may still be found in future excavations if they are located to the south. This example shows that objects starting from an origin are dispersed not only longitudinally (flow direction) but also laterally in the river with a larger transport distance. Certainly, effects at the edge of the flow channel (sweep water) and the observation that water tends to flow in a helical motion must also be considered. Nevertheless, the fundamental assumption remains that the lateral dispersion of transported elements increases with distance.

5.9 Biogenic bone modifications and possible producers

A large number of the recovered bone finds from Hammerschmiede show possible traces of biogenic (animal or plant / algae / bacterial) modification. The following is an overview of the traces found in the Hammerschmiede material and an assignment to a possible producer. Traces are mainly found on disarticulated, isolated specimens and have not yet been recognised on bones belonging to skeletal strewnfields of single vertebrate individuals. Related or associated finds possibly have not been available to the modifiers (temporally, spatially), which could possibly imply that the better-preserved skeletal find complexes arrived at the site rather late, shortly before the final sediment cover.

Splinter bones: A significant amount of bone finds show mechanical breakage and fracturing or splintering into smaller parts and fragments (Fig. 5.9.1). Many of these splinters and bone shaft pieces show carnivorous tooth marks (puncture marks) in addition to fracture patterns possibly resulting from chewing (Fernández-Jalvo & Andrews 2016, Arnold 2022). The way the bones are fragmented indicates that they were manipulated mechanically in a "fresh" state (e. g., Fernández-Jalvo & Andrews 2016). Many bone shaft splinters collected as isolated finds can be assigned in terms of size to the long bones of bovids, tragulids, cervids, or suids. The majority of these bone shaft fragments show typical splinter fractures and tooth marks possibly produced by larger carnivores. Since larger felids and barbourofelids possessed relatively slender canines and likely did not employ them for bone-crushing, it is improbable that they produced the observed fracture patterns. However, it cannot be excluded that they – or other feliforms – may have caused some damage using their carnassials, although their bone-processing capabilities are generally considered less efficient compared to those of hyaenas, amphicyonids, or large mustelids such as *Eomellivora*, which are more likely candidates for marrow scavenging. While there are Pleistocene sites such as Untermassfeld (Germany) where carnivore taxa could be clearly linked to specific bone modification patterns (e. g., Arnold 2022), the diversity of large carnivores present at Hammerschmiede – combined with the current lack of detailed taphonomic data on many of these taxa – precludes definitive taxonomic attribution at this stage. Overall, the angled fragmentation patterns with curved cleavage are quite similar to regurgitated and chewed bones from modern hyaenas as shown by Fernández-Jalvo & Andrews (2016).

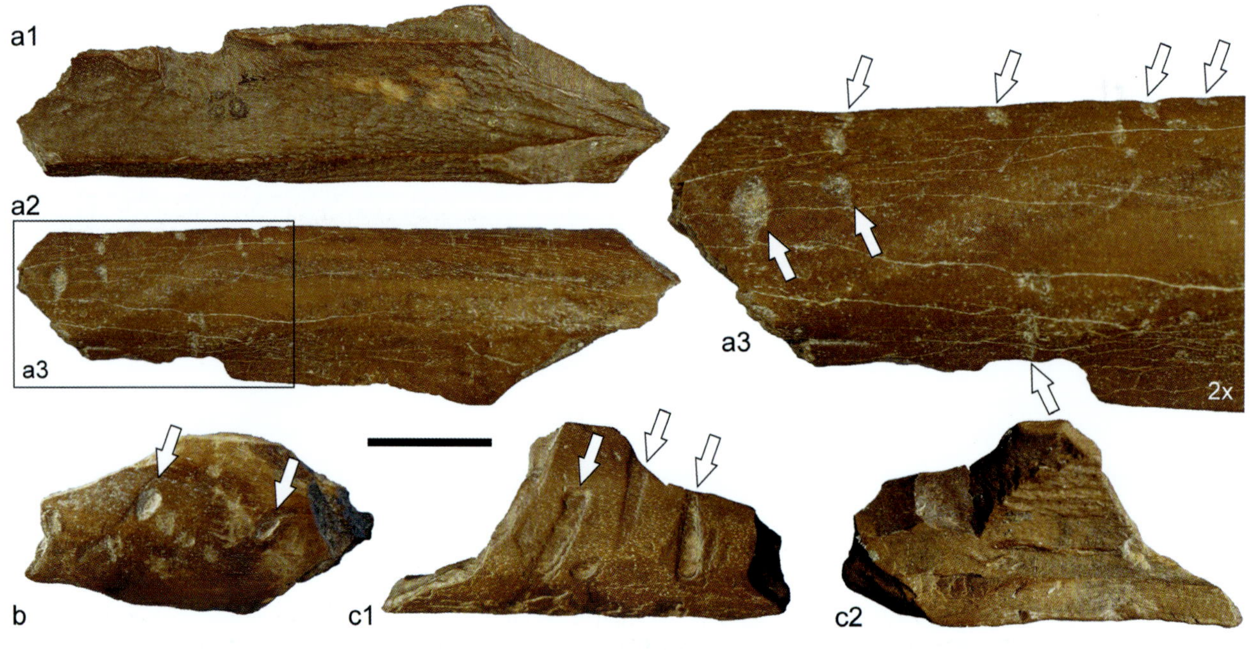

Figure 5.9.1. Bone splinters with tooth marks from the early Late Miocene Hammerschmiede deposits and the local stratigraphic levels HAM5 and HAM4. (a–c) Bone fragments most probably resembling diaphysis fragments of the antelope *Miotragocerus monacensis* showing surface pits (a2–3, b) and linear striations (b, c1) possibly produced by carnivore chewing. White arrows mark particularly conspicuous pits or linear striations. (a) GPIT/MA/19653; (b) GPIT/MA/19654; (c) GPIT/MA /16452. Scale bar equals 10 mm (a1–2, b, c) or 5 mm (a3).

Figure 5.9.2. Gnaw marks on bones from the early Late Miocene Hammerschmiede deposits and the local stratigraphic levels HAM5 and HAM4. (a) Plastron fragment of a Testudinidae turtle with linear marks possibly produced by rodent gnawing restricted to one side of the bone, in the overview (a1) and detail view (a2). (b) Mammalian diaphysis fragment with linear marks at one fracture edge. (c) Diaphysis fragment (humerus or femur) of the snapping turtle *Chelydropsis* sp. with very few linear marks with bite and counterbite possibly produced by gnawing of small rodents or other small mammals like Eulipotyphla, in the overview (c1) and detailed views (c2–3). (d) Avian diaphysis showing multiple linear striations possibly produced by gnawing of rodents or other small mammals, in the overview (d1) and detail view (d2). (e) Right proximal metatarsal II of the beaver *Steneofiber depereti* with very tiny linear marks with bite and counterbite striations possibly produced by gnawing of a very small rodent or another very small mammal, in overview (e1) and detail view (e2). (a) GPIT/RE/13656; (b) SNSB-BSPG 2020 XCIV-3102; (c) GPIT/RE/15229; (d) SNSB-BSPG 2020 XCIV-1015, (e) GPIT/MA/16621. Scale bar equals 10 mm (a1, b, c1, d1, e1) or 5 mm (a2, c2–3, d2, e2).

Figure 5.9.3. Insect marks on different bones from the early Late Miocene Hammerschmiede deposits and the local stratigraphic levels HAM5 and HAM4. (a–d) different bone types showing partially crisscrossing or intersecting linear or slightly curved marks possibly produced by insects or other arthropods using their mandibles. (a) Plastron fragment of a trionychid turtle showing linear marks on the flat interior surface of the carapace bone possibly produced by insects, in the overview (a1) and detailed views (a2–3). (b) Fragment of a snapping turtle (*Chelydropsis* sp.) scapula with linear marks on the surface possibly produced by arthropods such as insects, in the overview (b1) and detailed views (b2–4). (c) Vertebra zygapophysis fragment of a large mammal with intersecting and curved carving marks possibly produced by larger arthropods like termites, in overview (c1) and detail view (c2). (d) Mammalian diaphysis fragment with curved and intersecting marks possibly produced by larger arthropods like termites, in overview (d1) and detail view (d2). (a) GPIT/RE/15230; (b) GPIT/RE/13198; (c) SNSB-BSPG 2020 XCIV-4696; (d) GPIT/MA/19655. Scale bar equals 10 mm (a1, b1, c1, d1), 5 mm (a2, b2, c2, d2) or 2.5 mm (a3, b3–4).

Gnaw marks (vertebrates): Very few bones show gnaw marks on their margin possibly produced by the gnawing rodents or other small mammals like eulipotyphlans (Fig. 5.9.2). The found structures resemble quite well the typical parallel striations produced by rodents (compare to e.g., Fernández-Jalvo & Andrews 2016).

Arthropod bioerosion (small scale): In addition to previously described bone modifications by vertebrates, also possible invertebrate trace fossils are present. Rare, but probably quite inconspicuous and difficult to find, are very fine surface traces of very delicate parallel and intersecting furrows, a few mm long, which occur, or are best found mainly on smooth surfaces (e.g., interior turtle shell bone surfaces) (Fig. 5.9.3 a–b). These fine structures usually appear next to each other as small patches with similar patterns. It appears as if two adjacent columns made of more-or-less parallel furrows are arranged slightly across each other. This could indicate a mechanical movement in an opposing direction, as is the case with mouthparts of several arthropods (e.g., termites) where mandibles can cross over each other. The search for the producer of these traces is quite difficult, however, they possibly represent traces produced by arthropods like insects such as termites or ants. Accidental, non-biogenic scratching of the surface can be ruled out, as the traces appear at different bone sites with a similar pattern in a different orientation.

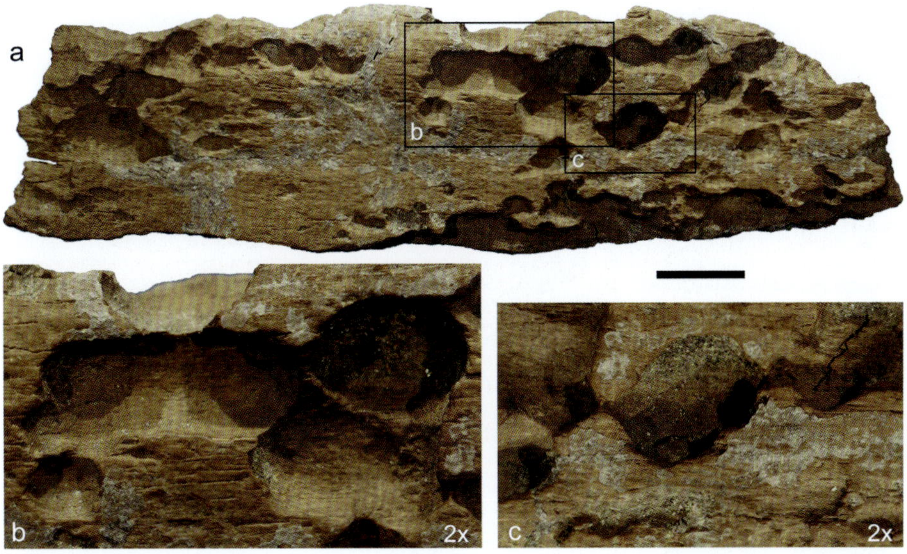

Figure 5.9.4. Insect marks on a large mammal bone from the early Late Miocene Hammerschmiede deposits and the local stratigraphic level HAM4. (a–c) Large fragment of a large mammal diaphysis (possibly of a Perissodactlya) showing circular and ovoid shaped depressions and furrows with partly undercut edges (overhanging) on the compacta surface of the bone, in the overview (a) and detailed views (b, c). These structures possibly represent insect marks and could probably resemble pupation cavities produced by dermestid beetles. (a–c) SNSB-BSPG 2020 XCIV-3416. Scale bar equals 10 mm (a) or 5 mm (b, c).

Arthropod bioerosion (large scale): Another rare pattern of most probable bioerosive origin, occurs mainly on flat bone areas and is characterised by an in general star-shaped, centralised and deeper imprint made of intersecting linear furrows that are only slightly larger than one cm at the maximum (Fig. 5.9.3 c–d). These markings show great similarity to trace fossils described from other vertebrate sites and could possibly be attributed to termites (compare e. g., Watson & Abbey 1986, Parkinson et al. 2010, Backwell et al. 2012, Fernández-Jalvo & Andrews 2016, Augustin et al. 2019).

Dermestid bioerosion: Undercut small hollow forms and surface modifications occur, especially on bones of large mammals of poor preservation quality (Fig. 5.9.4). These hollows and furrowed surface areas possibly could be assigned to dermestid beetles, and their feeding and carving activity building pupation chambers (compare e. g., Kitching 1980, Fernández-Jalvo & Andrews 2016, Keiler et al. 2020, Augustin et al. 2021).

Plant marks: A common feature on many isolated bones and bone fragments are linear, bifurcated, and anastomosing furrow marks caused by roots (Fig. 5.9.5). Especially root traces are marks that require longer duration time. Root marks indicate the embedding of bones into the rooted zone of soils for a certain period, where plants leach out the bones as phosphate supplier with their roots. Since in the Hammerschmiede deposits affected bones and perfectly preserved bones are found next to each other, these traces indicate a secondary relocation of the affected specimens by the fluvial processes.

Corrosion by aquatic vegetation: The bone preservation in the Hammerschmiede is extremely heterogeneous. In addition to bones of excellent surface quality, there are also many pieces with severe bone surface damage and structures reminiscent of corrosion, such as corrugated, furrowed, and irregularly roughened surfaces and surface areas of bones (Fig. 5.9.6) that could represent corrosion effects of "aquatic vegetation" (sensu e. g., Pesquero et al. 2010, Fernández-Jalvo & Andrews 2016). Some of these structures may be due to mechanical rolling, plant or animal traces. However, there are bone surfaces that are more suggestive of chemical dissolution processes. Since chemical bone corrosion would require pH values in the highly acidic or alkaline range, which would completely contradict the observations made in the deposits of high water quality (freshwater pearl mussels, aquatic gastropods, salmonid fishes, diverse amphibians), another source must be responsible for these modifications.

In the case of a suid mandible, it becomes clear that this form of surface modification was only effective under certain conditions. While the bone surface of the left hemimandible is of a higher quality of preservation, the other side of the jaw, as if separated by an imaginary line, shows a strong corrosive impact (Fig. 5.9.7).

A possible explanation for this is the assumption that some bones are placed over a longer period in one position, semi-buried in channel base sediments and partly exposed to flowing water at the river bottom. The left mandibular ramus was broken before embedding, and the hook-like symphysis may have helped to anchor the bone in the soft sediment. The orientation of the lower jaw in its find position aligns with the reconstructed flow direction from south (right) to north (left) (Fig. 5.9.7 d). Additionally, corrosion on the distal mandible suggests that typical flow turbulence, such as scour effects, may have influenced this area. In the light-flooded area of the flowing waters, algal or bacterial mats or water plants certainly settle on the exposed bone surfaces, which use the bone as a phosphate supplier source and hard ground in the surface

Figure 5.9.5. Root marks on different vertebrate bone and tooth fragments from the early Late Miocene Hammerschmiede deposits and the local stratigraphic levels HAM5 and HAM4. (a) Diaphysis fragment possibly belonging to the antelope *Miotragocerus monacensis* with intense linear marks most probably produced by plant root corrosion, in internal (a1), breaking edge (a2), and compacta surface (a3) views and a detailed view of the internal surface (a4). Note that root marks are found on both the internal and the external surface of the bone but not at the breaking edge (a2). This indicates that the bone did not break until after root penetration during a new redepositional process. (b) Peripheral bone fragment of a Geoemydidae swamp turtle (possibly *Mauremys* sp.) carapace with linear marks on the internal (b1) and external (b2) surface of the bone most probably produced by roots. (c) lower incisor of the small beaver *Euroxenomys minutus* with linear marks forming a dichotomous network structure produced by roots, in lateral (c1), anterior (c2), and detail anterior views (c3). Note, that the dentine part is densely corroded without showing any detailed root structures while the enamel clearly exhibits typical root mark network patterns by differences in colouring. (a) GPIT/MA/19656; (b) GPIT/RE/15231; (c) GPIT/MA/17322. Scale bar equals 10 mm (a1–3, b1–2, c1–2) or 5 mm (a4, c3).

area (compare e. g., Pesquero et al. 2010, Fernández-Jalvo & Andrews 2016). In addition to this, grazers such as snails might have fed on these biofilms together with the unstable bone surface and thus caused the massive irregular surface destruction as a recurring combination. Similar modifications are described by e. g., Pesquero et al. (2010) or Fernández-Jalvo & Andrews (2016) as corrosion by aquatic vegetation (moss, algae, or bacterial mats).

Rolling and polishing: In parallel with fragmentation, disarticulation, and biogenic surface modifications, many bones and bone fragments show surface polishing and rolling defects such as rounded corners and edges. These traces are probably mainly due to fluvial transport processes and prove a correspondingly greater transport distance for these finds than for "fresh" and undamaged bones.

Digestive corrosion: The teeth and jaws of small mammals, especially of the small beaver *Euroxenomys minutus*, show a different form of corrosion. In addition to an observed high degree of fragmentation of the material, several dental fragments of this small beaver species show enamel lesions on the exposed tooth regions, while the dentine and jawbone themselves are often little or not affected by corrosion (Lechner & Böhme 2023: figure 8). In the incisors the enamel is sometimes completely missing from the tip, while in the cheek teeth lesions occur at the border between the enamel and the jawbone, whereas other enamel areas show hardly any dissolution. In this recent study, Lechner & Böhme (2023) conclude that such lesions, often involving only the enamel, can most probably be attributed to digestive corrosion by birds of prey or mammalian carnivores. This probably indicates that the small beaver and many other small vertebrate finds from Hammerschmiede represent a predatory bone assemblage.

Coprolites: In addition to bones with direct traces, there are also large numbers of phosphatic coprolites found. Besides many morphological differences, two main size categories can be distinguished. There are large coprolites in the size range of several cm, which probably originate from hyaenas or other bone-cracking mammalian predators. Some of them also contain macroscopically recognisable bone fragments (Fig.

Figure 5.9.6. Bone surface corrosion on turtle bones from the early Late Miocene Hammerschmiede deposits and the local stratigraphic levels HAM5 and HAM4. (a) Plastron fragment of the trionychid turtle *Trionyx* sp. with preserved typical sculpture (bottom left) where a large part of the actual sculpture is missing (white dashed area) possibly corroded by bacteria or algae. (b) Epiplastron of the snapping turtle *Chelydropsis* sp. is possibly corroded, while small zones remain unaffected preserving a smooth surface, in the overview (b1) and a detail view (b2). (b2) shows irregularly distributed circular zones of initial corrosion that merge into larger patches, while directly adjacent areas remain unaffected – possibly caused by corrosive algal or bacterial mats. (c) internal view of a plastron fragment of the trionychid turtle *Trionyx* sp. showing very tiny dots and smaller circular areas of bone surface corrosion probably produced by aquatic vegetation (algal or bacterial corrosion), in the overview (c1) and detailed views (c2–3). (a) GPIT/RE/15143; (b) GPIT/RE/15232; (c) GPIT/RE/15233. Scale bar equals 10 mm (a, b1, c1) or 5 mm (b2, c2–3).

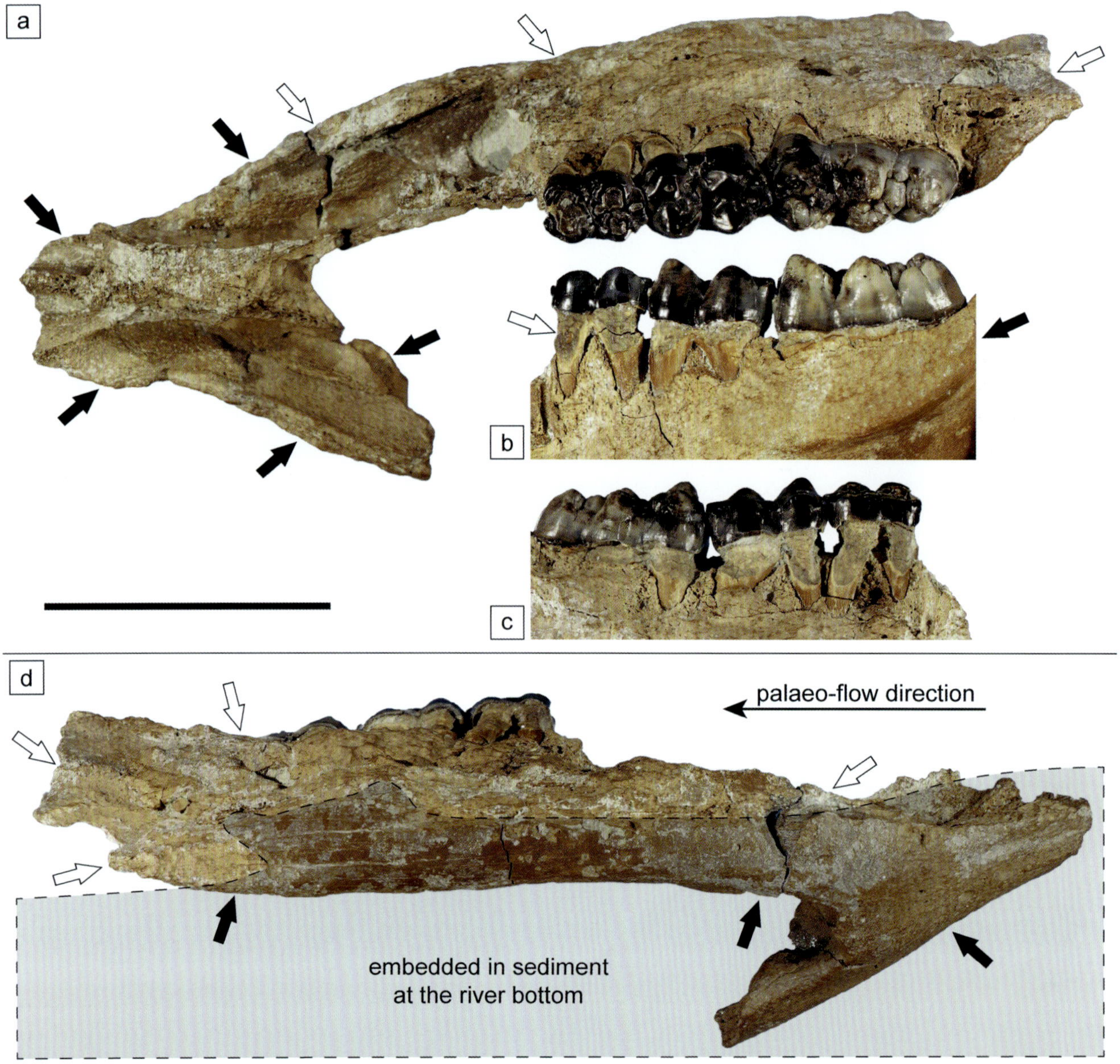

Figure 5.9.7. Bone surface corrosion on a suid mandible from the early Late Miocene Hammerschmiede deposits and the local stratigraphic level HAM4. (a–d) Fragmentary mandible of the suid *Propotamochoerus palaeochoerus* (SNSB-BSPG 2020 XCIV-9350) with right m1, m2, and m3 dentition in occlusal (a), lingual detail (b), buccal detail (c), and ventral (d) views. Black arrows indicate well-preserved bone surface while white arrows point to the very corroded surface areas. An imaginary line divides the corroded and well-preserved regions. This possibly demonstrates that the specimen was half-embedded in the sediment at the river bottom. The reconstructed flow direction from the right (south) indicates an anchoring effect of the broken symphysis. As a result, there was a division into a more protected half (embedded) and a counterpart that presumably was overgrown and corroded by aquatic vegetation (algae or bacterial mats) on the river bottom (d). Fractures and defects must have occurred before this embedding. Scale bar equals 5 cm.

5.9.8.1; further compare e.g., Fernández-Jalvo & Andrews 2016, Gross et al. 2023).

Furthermore, a very large number of small- and medium-sized and also micro-coprolites (<5 mm) are found (see section 5.7), often containing prey bones of mainly aquatic vertebrates (pharyngeal teeth of cyprinid fishes or fish vertebrae, vertebrae of proteid salamanders like *Mioproteus* sp.; Fig. 5.9.8.2; further compare e.g., Dentzien-Dias et al. 2021). The high disparity in size and morphology suggests that a multitude of predators are potential coprolite producers.

Aquatic insects (burrows): Not only the bones themselves but also the surrounding sediments sometimes show trace fossils. U-shaped inclined structures (burrows) in different parts of the clayey HAM4 channel base are assignable to larvae of borrowing mayflies (Ephemeroptera) colonising the bare river bottom whereby filtering the oxygenated water (Fig. 5.9.9; Fürsich & Mayr 1981, Bolliger 1999b, Knaust 2013, Uchmann et al 2016).

The U-shaped tube-like structures found in the clay are filled with fine sand typical of HAM4 sediments.

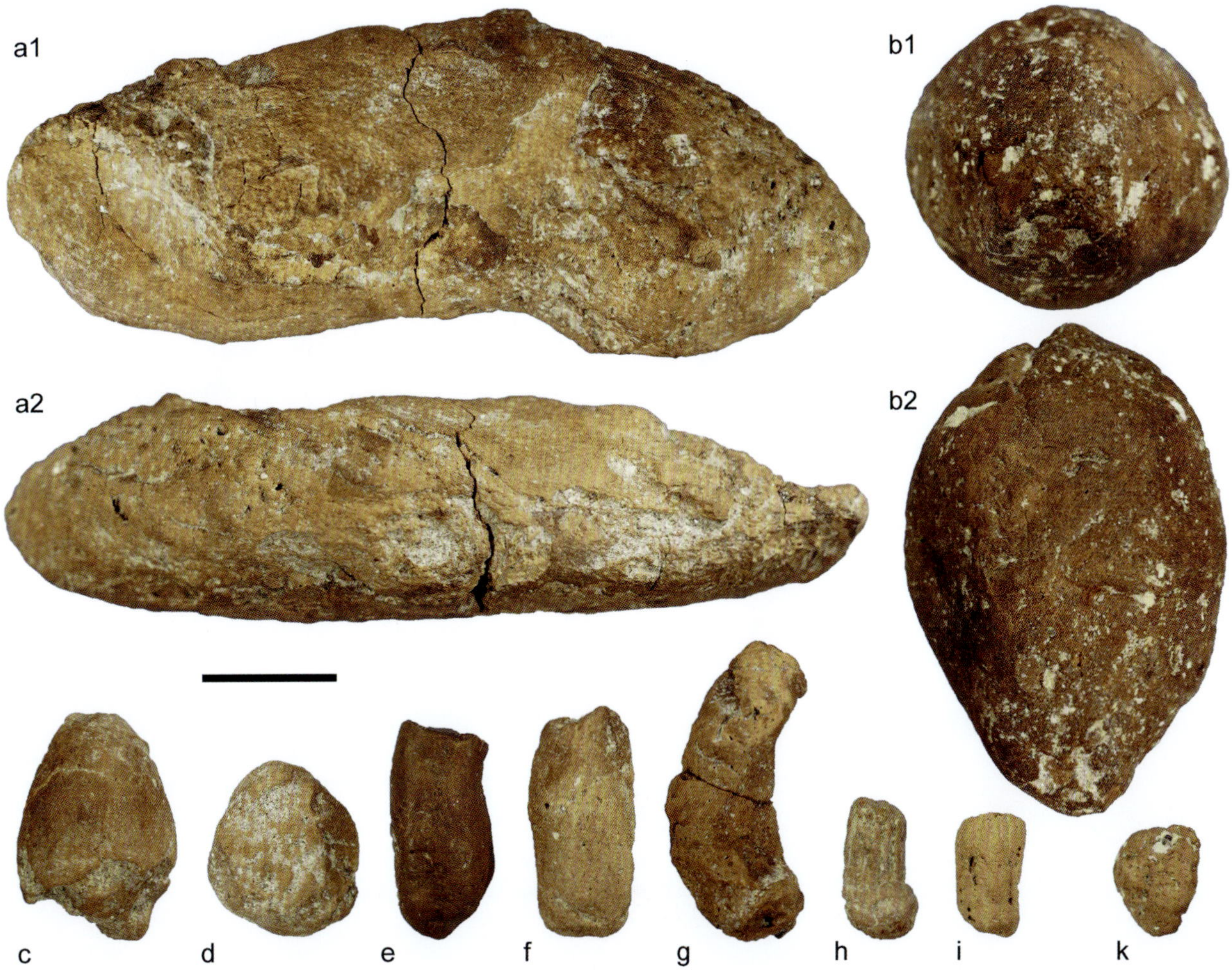

Figure 5.9.8.1. Coprolites from the early Late Miocene Hammerschmiede deposits and the local stratigraphic levels HAM5 and HAM4. (a–b) very large phosphatic coprolites of elongated (a) or more ovoid-shaped form (b) probably produced by larger mammalian carnivores such as hyaenids. (c–k) medium and small-sized phosphatic coprolites of ovoid (c), knobbly (d), oblong (e–g), striated oblong (h–i), or irregular flat (k) shapes. Due to the enormous diversity of shapes and sizes, various causative producers could certainly be considered. (a) SNSB-BSPG 2020 XCIV-3603; (b) GPIT/MA/10968; (c–k) GPIT/MA/12554. Scale bar equals 10 mm.

Figure 5.9.8.2. A small coprolite with recognizable object inclusion from the early Late Miocene Hammerschmiede deposits and the local stratigraphic level HAM4. (a1–2) Different views of a small coprolite containing an amphibian vertebra, possibly belonging to the proteid salamander *Mioproteus* sp. (SNSB-BSPG 2020 XCIV-3300). Scale bar equals 5 mm.

Figure 5.9.9. Inclined U-shaped ichnofossils attributed to *Rhizocorallium jenense* from the early Late Miocene Hammerschmiede deposits and the local stratigraphic level HAM4. The clay basement of the fine-sand-filled HAM4 channel shows in some areas U-shaped, fine-sand-filled structures (a1–2). These structures possibly represent burrows of larvae of mayflies (Ephemeroptera). (a2) reconstruction and sketch of the burrows visible in figure (a1). (b) *In situ* section of the burrows in the western part of the HAM 4 North channel base (2024), oriented transverse to the channel course (with the channel base dipping to the right). The U-shaped structures are arranged longitudinally to the channel course. (a) SNSB-BSPG 2020 XCIV-8915. Scale bars equal 10 mm.

Figure 5.9.10. Pathologies on turtle bones from the early Late Miocene Hammerschmiede deposits and the local stratigraphic levels HAM5 and HAM4. (a) Pygal and two peripheral bones of the chelydrid snapping turtle *Chelydropsis* sp. showing bone lesions in the form of dents and pits, in overview (a1) and detail view (a2). The edges of the damaged areas show an active bone reaction, which suggests a healing process of inflammation. (b) Xiphiplastron of a Geoemydidae swamp turtle (*Mauremys* sp.) showing bone lesions in the form of irregular dents in the bone surface, in the overview (b1) and detail view (b2). The cavity is not simply of mechanic origin, but the bone shows a reaction in the form of healing through bone remodeling and new formation. (a) GPIT/RE/13994; (b) GPIT/RE/13360. Scale bar equals 10 mm (a1, b1) or 5 mm (a2, b2).

Similar traces, described as *Rhizocorallium* (Bolliger 1999b) or *R. jenense* (Fürsich & Mayr 1981), commonly occur in the younger part of the Upper Freshwater Molasse of Switzerland and Bavaria and are also produced by several extant mayfly species, representing burrows of their larvae (e. g., Uchman et al. 2016). Larvae of Ephemeroptera represent filter-feeding aquatic organisms that are usually good indicators for unpolluted and clean rivers with a rich dissolved oxygen content (e. g., Iyagbaye et al. 2017 and citations therein). As a consequence, this would prove that there were also phases in the HAM4 sedimentation in which the bare clay bottom of the channel was without sediment cover for a certain period of time.

As a final point in this section, it should be noted that certain phenomena, which might initially appear to have a taphonomic origin, are more likely of a pathological nature (Fig. 5.9.10).

In several bone finds from the carapace of turtles, small lesions in the form of dents or holes occur on the surface. These bone surface damages are especially exhibited in the areas of dermal scute margins and borders. In contrast to the aforementioned taphonomic observations, in most of these bone lesions, the bone shows a reaction to the damage in the form of healing activity and remodelling of the bone which proves a process during a lifetime (Fig. 5.9.10). Although similar lesions have, in some cases, been confidently attributed to carnivore bite marks (e. g., Kahlke et al. 2015), such an interpretation appears unlikely here. A more complete set of material would be required to support this, and the presence of healing suggests that, if inflicted by a predator, the attack was not fatal. These lesions most likely represent a pathological condition, possibly comparable to necrotic shell disease in extant turtles, which is presumed to be primarily caused by bacterial infections, often in combination with fungal agents (e. g., Lovich et al. 1996).

6 Conclusion and outlook

The Hammerschmiede locality, known to science for over half a century, initially attracted attention primarily due to its taxonomic richness. Since the beginning of systematic excavations by the University of Tübingen in 2011 – and particularly with the implementation of advanced documentation and excavation methods from 2017 onwards – the quantity and quality of recovered fossils have significantly increased. These advances enhance spatial analyses and allow for deeper insights into sedimentary and biostratinomic processes.The sedimentary sequence of the Upper Series lithostratigraphic unit at Hammerschmiede exhibits a gentle northward dip of 0.5°–0.85°, indicating affiliation with the Inclined Foreland Molasse and the southern wing of the North Alpine Molasse Basin, which is tilted due to uplift of the Subalpine Molasse.

Seven fossil-bearing strata (HAM1–HAM7), chronologically dated to between 11.62 and 11.56 Ma, have been identified. While four of them represent historical outcrops (HAM1–3, 6), three (HAM4, 5, 7) remain accessible today, with ongoing excavations focusing on HAM4 and HAM5. The sedimentary facies within the 27 m thick exposed strata suggest a lower section composed predominantly of terrigenous floodplain deposits (containing the HAM5 level) and an upper section shaped by a dynamic meandering river system (HAM4), with associated channels, crevasse splays, and floodplain lakes (HAM1, 7).

The HAM4 channel displays a sinuous course from southwest to northeast, corroborated by both macro- and microscale flow direction analyses. These include orientations of elongated elements such as bones and wood, as well as the alignment of mass accumulations of freshwater pearl mussels. Taphonomic observations, including inferred strewnfields of individual skeletons, further support this reconstruction. Sedimentary structures in the HAM4 channel reveal a lateral and downflow accreting meandering river with four fining-upward point bar units. Fossil vertebrates, molluscs, and wood fragments are concentrated in middle to fine sands of the corresponding channel lag and the large-scale trough cross-stratified lower bar sediments. Cut-bank erosion is documented from mud clasts and reworked terrestrial gastropods and pedogenic carbonates, which are accumulated in the channel lag. The finer-grained inclined bedded sand with small-scale trough cross-stratifications of the upper bar facies contains no or very rare fossils and has been removed before excavation. The channels infill contains practically no pebble- or cobble-sized lithic clasts, suggesting that the source region of the HAM4 river was not the Alpine Orogen, but instead, its origin was intrabasinal, similar to today's nearby rivers Mindel and Günz. Further studies on heavy mineral assemblages should be conducted to test this hypothesis. The reconstructed dimensions of the HAM4 channel suggest that at bankfull stage, the river was up to 50 m wide and 4 m deep, while at low discharge, it narrowed to only 15 m in width and 1 m in depth.

Several skeletal strewnfields were detected at different excavation areas within lower bar deposits of HAM4, e.g., a suid, a tragulid, two antelopes, and avian remains. Furthermore, articulated finds of turtle shells and a bird leg (*Allgoviachen tortonica*, Mayr et al. 2022) were made. Several anatomically arranged skeletal elements indicate that at least connected parts (e.g., by ligaments) of carcasses were introduced to the fluvial system. This suggests partial carcass dehydration during the decomposition out of reach of scavengers and subsequent transportation and deposition of these elements by the river, potentially at higher discharge. The bone content within the channel lags is, similar to HAM5 (see below), very heterogeneous and mixed. Elements of assumed strewnfields of individuals are exceptions and most likely represent carcasses that were inserted late before the final burial.

The local stratigraphic level HAM5 most likely represented a smaller watercourse, more in the size of a rivulet and rather stable in place. The exact flow behaviour of the small HAM5 channel could not be clearly reconstructed due to the small extent of the available documented excavation area. Nevertheless, a slightly meandering course is suggested due to field investigations. The HAM5 flow direction towards the north is mainly estimated based on sedimentological evidence. The numerous reconstructed skeletal strewnfields of presumed single vertebrate individuals clearly indicate that a flow direction from south to north can be regarded as definite.

The HAM5 shows a main erosive channel depression, which was successively filled by two generations of fining upward sequences including basal channel lag deposits and on-top vegetation markers (roots) indicating slight temporal separations. These two channel fillings mostly enriched osteological specimens in channel lag deposits. Taxonomic, osteological, and preservative heterogeneity of the found material shows a strong intermixing and accumulation of objects from the closer but also more distant regions over a certain time, as many specimens are placed incoherently next to each other. Evidence that carcasses from the immediate vicinity washed into the HAM5 channel is provided by the reconstruction of a bony strewnfield representing a male individual of the great ape *Danuvius guggenmosi*. This discovery, as well as other skeletal strewnfields, prove that those carcasses were introduced into the area at a fairly late stage, shortly before the final burial, and thus represent autochthonous elements of the rivulet surrounding ecosystem. A third and final phase of sediments at the HAM5 level additionally overlies a lateral overbank area of the main channel, which is characterised by low contents of aquatic vertebrates and a different qualitative and quantitative fossil composition compared to the first two phases. It is shown that this lateral overbank deposit was subject to significantly less turbulent deposition and thus exhibits various detectable relations between finds. A variety of mammalian strewnfields – including a juvenile deinothere, an antelope, two tragulids, a beaver, and a large feliform carnivore – reveal a uniquely high concentration of carcasses at the Hammerschmiede site.

In general, the Hammerschmiede strewnfields provide an atypical pattern of bone distribution in fluvial channels. Previous assumptions predict that the distribution of the bones depends primarily on their density and surface properties as well as on the flow velocity (e.g., Voorhies 1969, Behrensmeyer 1975). The Hammerschmiede deposits contain accumulations of bones that are actually quite easy to transport (e.g., ribs) together with the heaviest and densest elements (e.g., teeth, skull), while most of the vertebrae and many other bones are missing. Here we hypothesised that this effect is produced by an influence of the soft-bottomed nature of the river channels, which adhesively sucked flat bones in particular, onto the bottom. More voluminous objects were possibly leveraged by the current and transported away. The requirement for this, however, is that the carcass (or its fragment) was introduced to the river and deposited in one piece, where it was first completely decomposed and disarticulated at rest so that all the bones could fall loose to the ground. This suggested that the HAM4 river was subject to strong fluctuations in currents and water discharge.

In addition to possible abiogenic abrasion (rolling) and fragmentation, many of the specimens found at HAM4 and HAM5 also exhibit a large number of biogenic traces. These prove that the bone materials in particular were decomposed and further utilized by larger carnivores, smaller vertebrates, and arthropods, as well as plants and possibly algae and bacteria. Root traces prove that some bones were embedded in rooted overbank soils over time until they were secondarily relocated. Arthropod (dermestids, termites, ants) traces as well as small mammal gnawing indicate skeletal remains have also decomposed under terrestrial conditions. An immense number of bone splinters prove that a major proportion of mammal carcasses was consumed and fragmented by larger, bone-cracking carnivores, and those remains were introduced and enriched within the river deposits. Similarly, suspected traces of digestive corrosion indicate that small mammals, such as the small beaver (*Euroxenomys minutus*) in particular, were preyed upon, and the finds largely represent digestive remains of e.g., birds of prey or carnivorous mammals.

Next to digestive corrosion, a second form of bone surface etching was detected on numerous fossil bones, which may be referred to as biofilm corrosion (or corrosion by aquatic vegetation). These traces possibly prove that bones were superimposed on or at least half embedded in the channel ground and exposed to the water column for a certain period of time. This hardground was possibly colonised by algae or bacteria, which superficially corroded and roughened the bones for the phosphatic suppliers.

Much more numerous than excavated larger fossils are microfossils. Here we investigated two test samples, one from HAM4 and one from HAM5, of together 3100 g of screen-washed and decalcified sediment, providing a total of 22,000 microfossils in the size from 1–5 mm. The analysis has revealed, beside a general similarity regarding the riverine faunal elements, some significant differences between both thanatocoenoses. In contrast to HAM5, the HAM4 channel contains a higher abundance of micro-charcoal and more lagomorphs, but fewer rodents and insectivore mammals, only a single (large size) flying squirrel species, a dominance of scincomorphs among the non-chelonian reptiles, and less numerous but larger phosphatic micro-coprolites. This led us to conclude that the riparian vegetation along the larger meandering river HAM4 experienced a higher frequency of wildfires, possibly creating more open space overbank, in addition to the wide sandy point bars at lower river discharge. This more open environment provided occupation space for a higher abundance of lizards and pikas and the more widely spaced trees were inhabited by the large flying squirrel *Albanensia grimmi*. Better sorting along the lower point bar is probably responsible for the comparatively lower abundance and larger size of micro-coprolites. A further test sample (205 g of sediment) from a potentially non- or lesser-sorted phosphatic micro-coprolite layer from HAM4 revealed an exceptionally high abundance (>10,000 coprolites) with a size range down to 200 µm. The accompanying vertebrate fauna consists exclusively of juvenile fishes, leading to the speculation that these micro-coprolites may represent bird faeces.

In contrast, the HAM5 environment was probably more densely vegetated due to lesser fire activity. Lagomorphs and scincomorphs were rare, but rodents and insectivore mammals were very common. The environment along the small rivulet with a denser (riparian) forest provided a habitat, not only for both syntopic hominids *Danuvius guggenmosi* and *Buronius manfredschmidi*, but also for the small flying squirrel *Blackia miocaenica* and the very common fossorial anomalomyid *Anomalomys gaudryi*. Phosphatic micro-coprolites are more numerous than in HAM4 but lesser sorted.

Our results from both fluvial levels fit well with already published mortality distributions of the large-sized beaver *Steneofiber depereti*, which differ significantly between the local stratigraphic levels of HAM5 and HAM4 (Lechner & Böhme 2022), in contrast to the small beaver *Euroxenomys minutus* (Lechner & Böhme 2023), which shows an almost identical mortality distribution in both layers and other environmentally different deposits (e.g., Rudabánya in Hungary). Using actualistic considerations, we could show that the mortality of the large beaver corresponds fairly closely to that of modern beavers (Lechner & Böhme 2022). The optimal habitat of *Steneofiber depereti* referred to larger rivers (such as HAM4) with residential family groups with multi-year parental investment. The HAM5 rivulet represented a rather inappropriate habitat, where transient prime adults migrated in search of their own territories. The small beaver, on the other hand, is ubiquitous (similar to the modern muskrat) and inhabits large rivers (e.g., HAM4), as well as smaller rivulets (e.g., HAM5) or swamp regions such as the locality Rudabánya, Hungary (Lechner & Böhme 2023). Due to its body size, mortality is controlled by external predation pressures rather than intraspecific parameters.

In addition to biostratinomic, ecologic, and partly necrologic parameters, also syn- and postgenetic and thus diagenetic influences on fossil preservation were analysed. HAM5 and HAM4 show clear differences in the preservability of different types of material. Diagenetic compaction is identified as one of the strongest impacts on deposited bones. Discrepancies between HAM4 and HAM5 are primarily due to the different sediment composition. The clay matrix of HAM5 strata show a high diagenetic compressibility (pore space reduction through water expulsion), and bones are more often compressed. In contrast, finds in the rather fine sandy HAM4 deposits were more easily preserved in three dimensions as the surrounding sediment is less compactable. The compression pressure was probably mainly caused by the sedimentary overburden, which could be assumed to be very high and only to a lesser extent by the glacial load during Pleistocene ice ages. In certain cases, modern compression and dislocation or shearing of bones by load and movement of mining vehicles as well as mining activities can not be ruled out. Finally, excavations are of course also responsible for damage to some finds or modification by chemical substances (e.g., adhesives).

Supplementary to mechanical preservation influences, geochemical observations have also been made at the Hammerschmiede deposits. High pyrite and uranium contents in HAM4 and HAM5 fossils indicate strong redox reactions in the subsurface of the channel sediments. This raise the question of whether this could be reconciled with the well-ventilated channel waters indicated by various aquatic faunal elements. It turns out that the sediments themselves were responsible for this, regardless of the water running over them. Presumably due to the high proportion of fine clastic sediments (mainly silt) and organic contents, the reduced pore water chemistry affects redox reactions during water infiltration. The sediment-water interface in the HAM4 river therefore represented a redox boundary. It remains as yet unexplained why the high iron-sulphide content in the wood becomes unstable after exhumation – causing it to disintegrate and making storage in the collection impossible – while the iron sulphide content in the bones remains stable. Future investigations could clarify whether this difference is possibly due to mineralogical variations and whether pyrite in bones and marcasite in wood was formed as iron sulphide and therefore explain why this is the case. The formation of marcasite and pyrite is pH-dependent: marcasite precipitates only under acidic conditions (pH <5; Murowchick & Barnes 1986), whereas both early and late diagenetic pyrite in bones forms under alkaline conditions, typically between pH 8 and 10 (Pfretzschner 1998, 2004).

Although no detailed investigation on mineral neoformation or iron sulphide mineralogy in bones from the Hammerschmiede locality has yet been conducted, several preliminary hypotheses can be proposed. The macroscopic pyrite crystals observed within the medullary cavities of some large mammal bones suggest formation during late diagenesis, when sulphide precipitation is governed by the surrounding sedimentary environment (Pfretzschner 2004). Such precipitation would require locally elevated pH, potentially buffered by phosphate released from the dissolution of bone apatite (Pfretzschner 2004) or derived from other sources, such as micro-coprolites or bird guano.

In contrast, the presence of potential marcasite in vascular plant fragments (e.g., tree trunks) from the HAM4 fluvial deposits suggests more acidic diagenetic conditions. These may result from organic acids released during microbial degradation of cellulose (e.g., Mali et al. 2019, Zhang et al. 2024). Furthermore, the near-complete decay of woody material in the rivulet-associated sediments of HAM5, combined with diagenetic aragonite dissolution (e.g., unionid shells), may indicate substantial shifts in pH between early and late diagenetic stages. Further research is clearly required to elucidate the diagenetic histories of the various fossil-bearing horizons at Hammerschmiede.

Finally, this report shows that comprehensive conclusions on fossil site genesis, taphonomy, and in particular biostratinomy, are only possible through proper and extensive documentation of the excavation. However, such techniques have not yet been widely adopted in palaeontology. Only a few sites such as Untermassfeld (Pleistocene; Germany, Thuringia) show a comparable comprehensive approach (e.g., Kahlke

2006, 2022). Particularly at sites with multiple superimposed fossil-bearing layers, small-scale investigations are invaluable for documenting intra-site variation over time. Fossil localities such as Sandelzhausen (Miocene; Bavaria, Germany) have been excavated in precisely this way, but the stratified distinctions have only been processed further in a few exceptions (e. g., Böhme 2010).

A deeper understanding of population dynamics – particularly with regard to mortality patterns and the minimum number of individuals (MNI) represented among large mammals – remains a key objective for future research at the Hammerschmiede locality. While preliminary observations suggest the potential for seasonal mortality signals, a conclusive assessment is not yet possible. Such interpretations require taxon-specific analyses that consider the presumed biology, behaviour, and ecological preferences of the respective species and groups. These aspects are currently being addressed in several ongoing doctoral projects and targeted faunal studies. In the future, more detailed investigations may yield insights into possible causes of death and taphonomic pathways, which, at present, remain largely unresolved.

The Hammerschmiede is a unique fossil site especially regarding its documented biodiversity. Out of the 169 known taxa, 151 belong to vertebrates among which are 86 mammalian species. So far 28 species of carnivores have been described, five of which represent aquatic fish predators (Kargopoulos et al. 2022). These numbers along with potentially up to nine sympatric turtle species, are virtually unrivalled in the fossil record and approach – or even exceed – present-day biodiversity levels in tropical bioregions. It has been suggested that the resources of the Hammerschmiede ecosystems must have been exceptionally rich, because the carnivore guild shows little niche overlap, thereby reducing competition, despite the high numbers of involved species (Kargopoulos et al. 2024b). For the aquatic ecosystem, this is certainly true. Here we estimate that 99.8% of the necromass in the riverine sediments of HAM4 belongs to juvenile fishes, accounting for up to 1.6 million specimens per metric ton of sediment.

Species richness at local scale (α-diversity) and between sites (β-diversity) are fundamental measures for biodiversity. Both are strongly influenced by abiogenic variables (habitat heterogeneity, climate, nutrient availability, disturbances), hence by geodiversity. The interdependence of bio- and geodiversity is controversial for extant ecosystems (Hawkins et al. 2003, Whittaker et al. 2007), but unexplored for the younger past. This is due to the lack of rich and well-excavated sites and the pervasive problem of time-averaging in the fossil record, preventing linkage of bio- and geodiversity on time scales below 10^4 years.

The Hammerschmiede excavation provides critical research perspectives for the future. As demonstrated, the point bar sequence of HAM4 resulted from lateral and downstream accretion processes of meander loops. In modern fluvial systems, such accretion typically occurs over timescales of only several hundred years (e. g., Micheli & Larson 2011). This is supported by sedimentological and stratigraphic characteristics at Hammerschmiede: the internal structure of HAM4 consists of well-bedded, laterally continuous, fine-grained point bar deposits lacking evidence for significant erosional unconformities or repeated avulsions, which suggests a relatively short-lived and uninterrupted depositional phase. Furthermore, each of the four distinguishable point bar bodies in HAM4 shows clear internal coherence and stacking patterns, indicating separate, successive meander accretion events rather than prolonged or amalgamated depositional periods. Based on analogues from active river systems and supported by the low variability of sedimentation rate as inferred from the lithology, each of these point bars is estimated to represent a time span of approximately 100–200 years. This allows for palaeobiologic and taphonomic analyses at a temporal resolution rarely attainable in the terrestrial fossil record, offering a unique opportunity to assess α-diversity patterns and ecological dynamics within tightly constrained time windows in the order of centuries. The Hammerschmiede is therefore able to bridge the disparity of the temporal scales of observation between past and extant ecosystems. Furthermore, the Hammerschmiede site provides abundant archives for geodiversity, which importantly also span temporal scales over several magnitudes, from millennia down to seasons. For example, pedogenic carbonate concretions integrate climatic signals (rainfall, soil temperature) over 10^3 years; aragonitic shells of the long-living freshwater pearl mussel *Margaritifera* (*Pseudunio*) grow for up to 200 years, which provides a research avenue for sclerochronology (Helama & Valovirta 2008); mammalian tooth enamel and cementum, but also incremental growth lines of the salmonid fish vertebra, document abiotic signals, and life history traits over several years; seasonal signals are recorded in the shells of terrestrial slugs or in the shells and opercula of aquatic mud snails. Fire severity can strongly influence biodiversity (Furnas et al. 2022), and changing habitat disturbances by fire is documented by different abundancies of micro- and macro-charcoal in HAM4 and HAM5. The Hammerschmiede site is, therefore, uniquely suited to reconstruct a highly resolved temporal evolution of vertebrate α-diversity in fluvial and riparian habitats under warmer-than-present climatic conditions and to investigate the contemporaneous biotic and abiotic traits across multiple temporal scales.

Acknowledgements

The authors sincerely thank I. Werneburg (GPIT) for granting access to the specimens housed in the Tübingen collection (GPIT and SNSB-BSPG stored at GPIT). We are also grateful to A.-C. Grupp, D. Wedekind (both SNSB-BSPG at Tübingen, Germany), and H. Stöhr (GPIT) for their meticulous preparation of the fossils. We thank A. Fatz and J. Francke (both Tübingen) for providing photographs of fossil material, and G. Ferreira and C. Kyriakouli (both 3D Imaging Lab at the Senckenberg Centre for Human Evolution and Palaeoenvironment at the Eberhard Karls University of Tübingen) for preparing surface scans and creating a 3D skeletal reconstruction.

The authors would like to express their sincere gratitude to the two private collectors, †S. Guggenmos (Dösingen) and M. Schmid (Marktoberdorf), for their excellent cooperation, as well as for generously transferring and making their finds and records available to us.

Our deepest appreciation goes to all participants of the numerous excavations at the Hammerschmiede fossil site since 2011, whose dedication and effort made the discovery and recovery of the studied material possible. We further acknowledge the invaluable contributions of the excavation team members from 2015 to 2024 (all GPIT or SNSB-BSPG): F. Augustin, A. Ayvazyan, C. Dietzel, H. El Atfy, J. Francke, N. Frank, J. Fuss, A.-C. Grupp, J. Hartung, P. Kampouridis, N. Kargopoulos, U. Kirscher, S. Kötter, C. Kyriakouli, M. Lex, T. Massonne, A. Matzke, M. Neuhof, S. Scheiben, H. Stöhr, A. Tröscher, D. Vasilyan, and D. Wedekind.

Our greatest gratitude goes to the Hartmann family, owners of the site, and the Geiger Group, the operating company of the Hammerschmiede clay pit, without whom excavations on this scale would not have been possible. We extend our special thanks to the site managers of past years, L. Saur, C. Heim, M. Hatt, S. Sachwitz and M. Zeberle as well as to the best machine operators on site, F. Becherer and J. Schmölz, whose expertise and support have been absolutely indispensable to the logistics of our excavations.

We extend our sincere gratitude to the UDO – Danuvius guggenmosi e. V. support association, whose board and members actively support our excavations and public outreach efforts.

Since 2020, the excavations and associated research have been generously supported by the Bavarian State Parliament, the Bavarian State Ministry of Research and the Arts, and the Bavarian Natural History Collections (Staatliche Naturwissenschaftliche Sammlungen Bayerns, SNSB). We are especially grateful to the parliamentary group of the *Freie Wähler* and in particular to Bernhard Pohl, Deputy Chairman and Member of the Bavarian State Parliament, for their dedicated support.

We also thank I. Gold-Lechner, W. Lechner, C. Dietzel, P. Kampouridis, F. Augustin, and A. Matzke for their valuable discussions and constructive feedback on the manuscript. Finally, we express our gratitude to editor U. Jansen and reviewers R.-D. Kahlke and M. Wuttke for their insightful comments and suggestions, which have greatly improved the manuscript. We further thank A. Lord for his linguistic improvements and I. Clasen for reviewing the figures.

References

Albrecht, P.K. (2017): Mineralogie der Knochen der Hammerschmiede. Unpublished BSc thesis, Eberhard Karls University Tübingen, Mathematisch-Naturwissenschaftliche Fakultät.

Allen, J.R.L. (1965): A review of the origin and characteristics of recent alluvial sediments. Sedimentology, 5: 89–191. https://doi.org/10.1111/j.1365-3091.1965.tb01561.x.

Arnold, J. (2022): Bovid bone modifications by Pachycrocuta from the Early Pleistocene site of Untermassfeld. – In: Kahlke R.-D. (Ed.): The Pleistocene of Untermassfeld near Meiningen (Thüringen, Germany). Part 5. Monographien des Römisch-Germanischen Zentralmuseums, Mainz: 40(5):X, 1423–1464.

Augustin, F.J., Matzke, A.T., Csiki-Sava, Z. & Pfretzschner, H.-U. (2019): Bioerosion on vertebrate remains from the Upper Cretaceous of the Haţeg Basin, Romania and its taphonomic implications. Palaeogeography, Palaeoclimatology, Palaeoecology, 534. https://doi.org/10.1016/j.palaeo.2019.109318.

Augustin, F.J., Matzke, A.T., Maisch, M.W. & Pfretzschner, H.-U. (2021): Dinosaur taphonomy of the Jurassic Shishugou Formation (Northern Junggar Basin, NW China) – insights from bioerosional trace fossils on bone. Ichnos 28(2): 87–96. https://doi.org/10.1080/10420940.2021.1890590.

Backwell, L.R., Parkinson, A.H., Roberts, E.M., d'Errico, F. & Huchet, J.-B. (2012): Criteria for identifying bone modification by termites in the fossil record. Palaeogeography, Palaeoclimatology, Palaeoecology, 337–338: 72–87. https://doi.org/10.1016/j.palaeo.2012.03.032.

Behrensmeyer, A.K. (1975): The Taphonomy and Paleoecology of Plio-Pleistocene Vertebrate Assemblages east of Lake Rudolf, Kenya. Bulletin of the Museum of Comparative Zoology, 146: 473–578.

Behrensmeyer, A.K. (1988): Vertebrate preservation in fluvial channels. Palaeogeography, Palaeoclimatology, Palaeoecology, 63(1–3): 183–199. https://doi.org/10.1016/0031-0182(88)90096-x.

Behrensmeyer, A.K. (1990): Transport/hydrodynamics of Bones. – In: Briggs, D.E.G. & Crowther, P.R. (Eds): Palaeobiology: A Synthesis. Blackwell Scientific Publications, Oxford, London, Edinburgh, Boston, Melbourne: 232–235.

Behrensmeyer, A.K. & Kidwell, S.M. (1985): Taphonomy's Contributions to Paleobiology. Paleobiology, 11(1): 105–119.

Böhme, M. (2002): Freshwater fishes from the Pannonian of the Vienna Basin with special reference to the locality Sandberg near Götzendorf, Lower Austria. Courier Forschungsinstitut Senckenberg, 237: 151–173.

Böhme, M. (2010): Ectothermic vertebrates (Actinopterygii, Allocaudata, Urodela, Anura, Crocodylia, Squamata) from the Miocene of Sandelzhausen (Germany, Bavaria) and their implications for environment reconstruction and palaeoclimate. Paläontologische Zeitschrift, 84: 3–41. https://doi.org/10.1007/s12542-010-0050-4.

Böhme, M. & Vasilyan D. (2014): Ectothermic vertebrates from the late Middle Miocene of Gratkorn (Austria, Styria). Palaeobiodiversity and Palaeoenvironments, 94: 21–40. https://doi.org/10.1007/s12549-013-0143-7.

Böhme, M., Ilg, A. & Ossig, A. (2006): New method to estimate paleoprecipitation using fossil amphibians and reptiles and the middle and late Miocene precipitation gradients in Europe. Geology, 34(6): 425–428. https://doi.org/10.1130/G22460.1.

Böhme, M., Ilg, A. & Winklhofer, M. (2008): Late Miocene "washhouse" climate in Europe. Earth and Planetary Science Letters, 275: 393–401. https://doi.org/10.1016/j.epsl.2008.09.011.

Böhme, M., Winklhofer, M. & Ilg, A. (2011): Miocene precipitation in Europe: Temporal trends and spatial gradients. Palaeogeography, Palaeoclimatology, Palaeoecology, 304: 212–218. https://doi.org/10.1016/j.palaeo.2010.09.028.

Böhme, M., Vasilyan, D. & Winklhofer, M. (2012): Habitat tracking, range dynamics and palaeoclimatic significance of Eurasian giant salamanders (Cryptobranchidae) – indications for elevated Central Asian humidity during Cenozoic global warm periods. Palaeogeography, Palaeoclimatology, Palaeoecology, 342–343: 64–72. https://doi.org/10.1016/j.palaeo.2012.04.032.

Böhme, M. & Vasilyan D. (2014): Ectothermic vertebrates from the late Middle Miocene of Gratkorn (Austria, Styria). Palaeobiodiversity and Palaeoenvironments, 94: 21–40. https://doi.org/10.1007/s12549-013-0143-7.

Böhme, M., Spassov, N., Fuss, J., Tröscher, A., Deane, A.S., Prieto, J., Kirscher, U., Lechner, T. & Begun, D.R. (2019a): A new Miocene ape and locomotion in the ancestor of great apes and humans. Nature, 575: 489–493. https://doi.org/10.1038/s41586-019-1731-0.

Böhme, M., Braun, R. & Breier, F. (2019b): Wie wir Menschen wurden: Eine kriminalistische Spurensuche nach den Ursprüngen der Menschheit – Spektakuläre Funde im Alpenraum. Heyne Verlag, München: 336pp.

Böhme, M., Spassov, N., DeSilva, J.M. & Begun, D.R. (2020): Reply to: Reevaluating bipedalism in *Danuvius*. Nature, 586: E4–E5. https://doi.org/10.1038/s41586-020-2737-3.

Böhme, M., Begun, D.R., Holmes, A.C., Lechner, T. & Ferreira, G. (2024): *Buronius manfredschmidi* – A new small hominid from the early late Miocene of Hammerschmiede (Bavaria, Germany). PLOS ONE, 19, e0301002. https://doi.org/10.1371/journal.pone.0301002.

Bolliger, T. (1999a): Family Anomalomyidae. – In: Rössner, G.E. & Heissig, K. (Eds): The Miocene Land Mammals of Europe. Verlag Dr. Friedrich Pfeil, Munich: 411–420.

Bolliger, T. (1999b): Trace fossils and trackways in the Upper Freshwater Molasse of central and eastern Switzerland. Neues Jahrbuch für Geologie und Paläontologie, Abhandlungen, 214: 515–536.

Boon, A. van der, Beniest, A., Ciurej, A., Gaździcka, E., Grothe, A., Sachsenhofer, R.F., Langereis, C.G. & Krijgsman, W. (2018): The Eocene–Oligocene transition in the North Alpine Foreland Basin and subsequent closure of a Paratethys gateway. Global and Planetary Change, 162: 101–119. https://doi.org/10.1016/j.gloplacha.2017.12.009.

Bruijn, H. de (1999): Superfamily Sciuroidea. – In: Rössner, G.E. & Heissig, K. (Eds): The Miocene Land Mammals of Europe. Verlag Dr. Friedrich Pfeil, Munich: 271–280.

Bruijn, H. de, Daams, R., Daxner-Höck, G., Fahlbusch, V., Ginsburg, L., Mein, P., Morales, J., Heinzmann, E., Mayhew, D.F., van der Meulen, A.J. et al. (1992): Report of the RCMNS working group on fossil mammals, Reisenburg 1990. Newsletters on Stratigraphy, 26(2/3): 65–118.

Daams, R. (1999): Family Gliridae. – In: Rössner, G.E. & Heissig, K. (Eds): The Miocene Land Mammals of Europe. Verlag Dr. Friedrich Pfeil, Munich: 301–318.

Dam, J.A. van (2010): Anourosoricini (Soricidae, Mammalia): new taxa and systematic position in the light of molecular phylogenetics. Journal of Vertebrate Paleontology, 30: 1221–1228.

Dentzien-Dias, P., Hunt, A.P., Lucas, S.G., Francischini, H. & Gulotta, M. (2021): Coprolites from shallow marine deposits of the Nanjemoy Formation, Lower Eocene of Virginia, USA. Lethaia, 54: 26–39. https://doi.org/10.1111/let.12380.

Doppler, G. (1989): Zur Stratigraphie der nördlichen Vorlandmolasse in Bayerisch-Schwaben. Geologica Bavarica, 94: 83–133.

Doppler, G., Heissig, K. & Reichenbacher, B. (2005): Die Gliederung des Tertiärs im süddeutschen Molassebecken. Newsletters on Stratigraphy, 41(1–3): 359–375. https://doi.org/10.1127/0078-0421/2005/0041-0359.

Efremov, I.A. (1940): Taphonomy: a new branch of paleontology. Pan-American Geologist, 74: 81–93.

Etter, W. (1994): Palökologie: eine methodische Einführung. Birkhäuser Verlag, Basel, Boston, Berlin: 294pp.

Fahlbusch, V. (1975): Die Eomyiden (Rodentia, Mammalia) aus der Oberen Süßwasser-Molasse Bayerns. Mitteilungen der Bayerischen Staatssammlung für Paläontologie und historische Geologie, 15: 63–90.

Fahlbusch, V. & Mayr, H. (1975): Microtoide Cricetiden

(Mammalia, Rodentia) aus der Oberen Süßwasser-Molasse Bayerns. Paläontologische Zeitschrift, 49: 78–93.

Fejfar, O. (1999): Microtoid Cricetids. – In: Rössner, G.E. & Heissig, K. (Eds): The Miocene Land Mammals of Europe. Verlag Dr. Friedrich Pfeil, Munich: 365–373.

Fernández-Jalvo, Y. & Andrews, P. (2016): Atlas of Taphonomic Identifications: 1001+ Images of Fossil and Recent Mammal Bone Modification, Vertebrate Paleobiology and Paleoanthropology. Springer Netherlands, Dordrecht: 368pp. https://doi.org/10.1007/978-94-017-7432-1.

Frison, G.C. & Todd, L.C. (1986): The Colby mammoth site: taphonomy and archaeology of a Clovis kill in northern Wyoming. University of New Mexico Press, Albuquerque: 238pp.

Fürsich, F.T. & Mayr, H. (1981): Non-marine *Rhizocorallium* (trace fossil) from the Upper Freshwater Molasse (Upper Miocene) of southern Germany. Neues Jahrbuch für Geologie und Paläontologie – Monatshefte, 1981: 321–333. https://doi.org/10.1127/njgpm/1981/1981/321.

Furnas, B.J., Goldstein, B.R. & Figura, P.J. (2022): Intermediate fire severity diversity promotes richness of forest carnivores in California. Diversity and Distributions, 28: 493–505. https://doi.org/10.1111/ddi.13374.

Fuss, J., Prieto, J. & Böhme, M. (2015): Revision of the boselaphin bovid *Miotragocerus monacensis* Stromer, 1928 (Mammalia, Bovidae) at the Middle to Late Miocene transition in Central Europe. Neues Jahrbuch für Geologie und Paläontologie – Abhandlungen, 276: 229–265. https://doi.org/10.1127/njgpa/2015/0481.

Gregor, H.-J. (1982): Die jungtertiären Floren Süddeutschlands: Paläokarpologie, Phytostratigraphie, Paläoökologie, Paläoklimatologie. I. Enke, Stuttgart: 278pp.

Gross, M., Prieto, J., Grímsson, F. & Bojar, H.-P. (2023): Hyena and 'false' sabre-toothed cat coprolites from the late Middle Miocene of south-eastern Austria. Historical Biology, 1–20. https://doi.org/10.1080/08912963.2023.2237979.

Gümbel, C.W. von (1861): Geognostische Beschreibung des bayerischen Alpengebirges und seines Vorlandes. Justus Perthes, Gotha: 950pp.

Hanson, C.B. (1980): Fluvial taphonomic processes: models and experiments. – In: Behrensmeyer, A.K. & Hill, A.P. (Eds): Fossils in the Making: Vertebrate Taphonomy and Paleoecology. University of Chicago Press, Chicago: 156–181.

Hartung, J. & Böhme, M. (2022): Unexpected cranial sexual dimorphism in the tragulid Dorcatherium naui based on material from the middle to late Miocene localities of Eppelsheim and Hammerschmiede (Germany). PLOS ONE, 33. https://doi.org/10.1371/journal.pone.0267951.

Hartung, J., Lechner, T. & Böhme, M. (2020): New cranial material of *Miotragocerus monacensis* (Mammalia: Bovidae) from the late Miocene hominid locality Hammerschmiede (Germany). Neues Jahrbuch für Geologie und Paläontologie – Abhandlungen, 298: 269–284. https://doi.org/10.1127/njgpa/2020/0948.

Hawkins, B.A., Field, R., Cornell, H.V., Currie, D.J., Guégan, J.-F., Kaufman, D.M., Kerr, J.T., Mittelbach, G.G., Oberdorff, T., O'Brien, E.M., et al. (2003): Energy, water, and broad-scale geographic patterns of species richness. Ecology, 84: 3105–3117. https://doi.org/10.1890/03-8006.

Helama, S. & Valovirta, I. (2008): The oldest recorded animal in Finland: ontogenetic age and growth in *Margaritifera margaritifera* (L. 1758) based on internal shell increments. Memoranda – Societatis pro Fauna et Flora Fennica, 84: 20–30.

Hír, J. (2005): Collimys dobosi n. sp. (Cricetidae, Mammalia) from the Late Astaracian (MN 8) vertebrate fauna of Felsötárkány 3/2 (Northern Hungary). Fragmenta Palaeontologica Hungarica, 23: 5–18.

Hugueney, M. (1999): Family Castoridae. – In: Rössner, G.E. & Heissig, K. (Eds): The Miocene Land Mammals of Europe. Verlag Dr. Friedrich Pfeil, Munich: 281–300.

Iyagbaye, L., Iyagbaye, R. & Omoigberale, M. (2017): Mayflies (Order Ephemeroptera) distribution as indicators of the water quality status of a stretch of Ovia River (Iguoriakhi), Edo State, Southern Nigeria. International Journal of Innovative Research & Development, 6. https://doi.org/10.24940/ijird/2017/v6/i9/SEP17041.

Jung, W. & Mayr, H. (1980): Neuere Befunde zur Biostratigraphie der Oberen Süßwassermolasse Süddeutschlands und ihre palökologische Deutung. Mitteilungen der Bayerischen Staatssammlung für Paläontologie und historische Geologie, 20: 159–173.

Kahlke, R.-D. (2006): Untermassfeld: a late early Pleistocene (Epivillafranchian) fossil site near Meiningen (Thuringia, Germany) and its position in the development of the European mammal fauna. BAR International Series 1578, Archaeopress, Oxford: 141pp.

Kahlke, R.-D. (2022): The Pleistocene of Untermassfeld – A synopsis of site origin, palaeobiodiversity, taphonomic characteristics, palaeoenvironment, chronostratigraphy, and significance in western Palaearctic faunal history. – In: The Pleistocene of Untermassfeld near Meiningen (Thüringen, Germany). Part 5: Monographien des Römisch-Germanischen Zentralmuseums, Mainz: 40(5): 1635–1734 + foldout II+III.

Kahlke, R.-D., Fritz, U. & Kierdorf, U. (2015): Badger (*Meles meles*) predation on European pond turtle (Emys orbicularis) during the Eemian interglacial (MIS 5e). Palaeobiodiversity and Palaeoenvironments 95(2): 223–235. https://doi.org/10.1007/s12549-014-0170-z.

Kälin, D. & Engesser, B. (2001): Die jungmiozäne Säugertierfauna vom Nebelbergweg bei Nunningen (Kanton Solothurn, Schweiz). Schweizerische Paläontologische Abhandlungen, 121: 1–61.

Kampouridis, P., Hartung, J., Lechner, T.S., Kargopoulos, N. & Böhme, M. (2024): Disparate occurrences of a chalicotheriine and a schizotheriine chalicothere (Mammalia, Chalicotheriidae) at the Late Miocene hominid locality Hammerschmiede (Germany). Paläontologische Zeitschrift, 98: 313–329. https://doi.org/10.1007/s12542-024-00685-x.

Kargopoulos, N., Kampouridis, P., Lechner, T. & Böhme, M. (2021a): A review of *Semigenetta* (Viverridae, Carnivora) from the Miocene of Eurasia based on material from the hominid locality of Hammerschmiede (Germany). Geobios, 69: 25–36. https://doi.org/10.1016/j.geobios.2021.07.001.

Kargopoulos, N., Kampouridis, P., Lechner, T. & Böhme, M. (2021b): Hyaenidae (Carnivora) from the Late Miocene hominid locality of Hammerschmiede (Bavaria, Germany). Historical Biology, 1–10. https://doi.org/10.1080/08912963.2021.2010193.

Kargopoulos, N., Valenciano, A., Kampouridis, P., Lechner, T. & Böhme, M. (2021c): New early late Miocene species of *Vishnuonyx* (Carnivora, Lutrinae) from the hominid locality of Hammerschmiede, Bavaria, Germany. Journal of Vertebrate Paleontology, 41: e1948858. https://doi.org/10.1080/02724634.2021.1948858.

Kargopoulos, N., Valenciano, A., Abella, J., Kampouridis, P., Lechner, T. & Böhme, M. (2022): The exceptionally high diversity of small carnivorans from the Late Miocene hominid locality of Hammerschmiede (Bavaria, Germany). PLOS ONE, 64. https://doi.org/10.1371/journal.pone.0268968.

Kargopoulos, N., Abella, J., Daasch, A., Kaiser, T., Kampouridis, P., Lechner, T. & Böhme, M. (2024a): The primitive giant panda *Kretzoiarctos beatrix* (Carnivora, Ursidae) from the hominid locality of Hammerschmiede: dietary implications. Papers in Palaeontology, 10: e1588. https://doi.org/10.1002/spp2.1588.

Kargopoulos, N., Valenciano, A., Abella, J., Morlo, M., Konidaris, G.E., Kampouridis, P., Lechner, T. & Böhme, M. (2024b): The carnivoran guilds from the Late Miocene hominid locality of Hammerschmiede (Bavaria, Germany). Geobios, S0016699524000445. https://doi.org/10.1016/j.geobios.2024.02.003.

Keiler, J.-A., Benecke, M. & Keiler, J. (2020): Bone modifications by insects from the early Pleistocene site of Untermassfeld. – In: Kahlke R.-D. (Ed.): The Pleistocene of Untermassfeld near Meiningen (Thüringen, Germany). Part 4. Monographien des Römisch-Germanischen Zentralmuseums, Mainz: 40(4): 1117–1132.

Kirscher, U., Prieto, J., Bachtadse, V., Abdul Aziz, H., Doppler, G., Hagmaier, M. & Böhme, M. (2016): A biochronologic tie-point for the base of the Tortonian stage in Europe terrestrial settings: Magnetostratigraphy of the topmost Upper Freshwater Molasse sediments of the North Alpine Foreland Basin in Bavaria (Germany). Newsletters on Stratigraphy, 49: 445–467. https://doi.org/10.1127/nos/2016/0288.

Kitching, J.W. (1980): On some fossil arthropoda from the Limeworks. Makapansgat, Potgietersrus. Palaeontologia Africana, 23: 63–68.

Kivell, T.L. (2019): Fossil ape hints at how walking on two feet evolved. Nature, 575: 445–446. https://doi.org/10.1038/d41586-019-03347-0.

Klein, S. (1937): Der Oberste Feinflinz im Alpenvorland und südlichen Tertiärhügelland. Zeitschrift der Deutschen Geologischen Gesellschaft, 89: 384–409.

Klein, S. (1938): Die Ausformung der Münchner Ebene durch fluvioglaziale Erosion jungtertiärer Schichten. Zentralblatt für Mineralogie, Geologie und Paläontologie, Abteilung B: Geologie und Paläontologie, 1938: 279–299.

Klein, S. (1939): Die miocän–pliocänen Grenzschichten nördlich von München. Zentralblatt für Mineralogie, Geologie und Paläontologie, Abteilung B: Geologie und Paläontologie, 1939: 278–292.

Klembara, J., Böhme, M. & Rummel, M. (2010): Revision of the anguine lizard *Pseudopus laurillardi* (Squamata, Anguidae) from the Miocene of Europe, with comments on paleoecology. Journal of Paleontology, 84(2): 159–196.

Knaust, D. (2013): The ichnogenus *Rhizocorallium*: Classification, trace makers, palaeoenvironments and evolution. *Earth-Science Reviews,* 126: 1–47.

Konidaris, G.E., Lechner, T., Kampouridis, P. & Böhme, M. (2023): *Deinotherium levius* and *Tetralophodon longirostris* (Proboscidea, Mammalia) from the Late Miocene hominid locality Hammerschmiede (Bavaria, Germany), and their biostratigraphic significance for the terrestrial faunas of the European Miocene. Journal of Mammalian Evolution: 1–39. https://doi.org/10.1007/s10914-023-09683-3.

Kordiuk, B. (1938): Zur Entwicklung des subalpinen Molassetroges. Abhandlungen der Preußischen Geologischen Landesanstalt, Neue Folge 187: 3–47.

Kretzoi, M. & Fejfar, O. (2005): Sciurids and Cricetids (Mammalia, Rodentia) from Rudabánya. Paleontographia Italica, 90: 113–148.

Kreutzer, L.A. (1988): Megafaunal butchering at Lubbock Lake, Texas: a taphonomic reanalysis. Quarternary Research, 30: 221–231. https://doi.org/10.1016/0033-5894(88)90026-9.

Kuhlemann, J. (2007): Paleogeographic and paleotopographic evolution of the Swiss and Eastern Alps since the Oligocene. Global and Planetary Change, 58: 224–236.

Kuhlemann, J., Frisch, W., Székely, B., Dunkl, I. & Kázmér, M. (2002): Post-collisional sediment budget history of the Alps: tectonic versus climatic control. International Journal of Earth Sciences, 91: 818–837.

Laatsch, W. (1931): Die Biostratinomie der Ganoidfische des Kupferschiefers. Palaeobiologica, 4: 175–238.

Lechner, T. & Böhme, M. (2022): The beaver *Steneofiber depereti* from the lower upper Miocene hominid locality Hammerschmiede and remarks on its ecology. Acta Palaeontologica Polonica, 67(4): 807–826. https://doi.org/10.4202/app.00997.2022.

Lechner, T. & Böhme, M. (2023): The largest record of the minute beaver *Euroxenomys minutus* (Mammalia, Castoridae) from the early Late Miocene hominid locality Hammerschmiede (Bavaria, Southern Germany) and palaeoecological considerations. Historical Biology, 36(7): 1415–1430. https://doi.org/10.1080/08912963.2023.2215236.

Lemcke, K. (1988): Das bayerische Alpenvorland vor der Eiszeit. Geologie von Bayern 1. Schweizerbart, Stuttgart: 175pp.

Lovich, E., Gotte, W., Ernst, H., Harshbarger, C., Laemmerzahl, F. & Whitfield, J. (1996): Prevalence and Histopathology of shell disease in turtles from lake Blackshear, Georgia. Journal of Wildlife Disease, 32(2): 259–265.

Lyman, R.L. (1994): *Vertebrate taphonomy,* Cambridge manuals in archaeology. Cambridge University Press, Cambridge [England], New York: 524pp.

Mali, T., Mäki, M., Hellén, H., Heinonsalo, J., Bäck, J. & Lundell, T. (2019): Decomposition of spruce wood and release of volatile organic compounds depend on decay type, fungal interactions and enzyme production patterns. FEMS Microbiology Ecology, 95(9), fiz135. https://doi.org/10.1093/femsec/fiz135.

Mayr, G., Lechner, T. & Böhme, M. (2020a): A skull of a very large crane from the late Miocene of Southern Germany, with notes on the phylogenetic interrelationships of extant Gruinae. Journal of Ornithology, 161: 923–933. https://doi.org/10.1007/s10336-020-01799-0.

Mayr, G., Lechner, T. & Böhme, M. (2020b): The large-sized darter *Anhinga pannonica* (Aves, Anhingidae) from the late Miocene hominid Hammerschmiede locality in Southern Germany. PlosOne, 15(5): 1–19. https://doi.org/10.1371/journal.pone.0232179.

Mayr, G., Lechner, T. & Böhme, M. (2022): Nearly complete leg of an unusual, shelduck-sized anseriform bird from the earliest late Miocene hominid locality Hammerschmiede (Germany). Historical Biology: 1–10. https://doi.org/10.1080/08912963.2022.2045285.

Mayr, H. & Fahlbusch, V. (1975): Eine unterpliozäne Kleinsäugerfauna aus der Oberen Süßwasser-Molasse Bayerns. Mitteilungen der Bayerischen Staatssammlung für Paläontologie und historische Geologie 15: 91–111.

Mein, P. (1967): Détermination de l'humerus de quelques Cricetidae fossiles. Problèmes actuels de paléontologie (Evolution des Vertébrés). Colloque International du Centre National de Recherche Scientifique, Paris: 515–521.

Mein, P. (1989): Updating of MN zones. – In: E. Lindsay, V. Fahlbusch & P. Mein (Eds): European Neogene Mammal Chronology. NATO ASI Series, 180: 73–90.

Meyer, B.L. (1956): Mikrofloristische Untersuchungen an jungtertiären Braunkohlen im östlichen Bayern. Geologica Bavarica, 25: 100–128.

Micheli, E.R. & Larsen, E.W. (2011): River channel cutoff dynamics, Sacramento River, California, USA. River Research and Applications, 27: 328–344. https://doi.org/10.1002/rra.1360.

Murowchick, J.B. & Barnes, H.L. (1986): Marcasite precipitation from hydrothermal solutions. Geochimica et Cosmochimica Acta, 50(12): 2615–2629. https://doi.org/10.1016/0016-7037(86)90214-0.

Müller, A.H. (1951): Grundlagen der Biostratinomie. Deutsche Akademie der Wissenschaften zu Berlin, Abhandlungen, 1950 (No. 3): 1–147.

Müller, A.H. (1963): Lehrbuch der Paläozoologie, Band 1: Allgemeine Grundlagen. G. Fischer Verlag, Jena: 387pp.

Nesin, V. & Kovalchuk, O. (2021): A new late Miocene *Anomalomys* species from western Ukraine with implications for the diversity and evolution of anomalomyid rodents in Eastern Europe. Historical Biology, 33(9): 1809–1816. https://doi.org/10.1080/08912963.2020.1742711.

Parkinson, A.H., Backwell, L.R., Roberts, E.M., D'Errico, F., Huchet, J.-B. & Val, A. (2010): Effects of termites on mammal and bird bone. – In: Mostovski, M.B., Ovechkina, M.N. (Eds), Proceedings of the 16th Conference of the Palaeontological Society of Southern Africa. Pietermaritzburg, South Africa: 79pp.

Pesquero, M.D., Ascaso, C., Alcalá, L. & Fernández-Jalvo, Y. (2010): A new taphonomic bioerosion in a Miocene lakeshore environment. Palaeogeography, Palaeoclimatology, Palaeoecology, 295: 192–198. https://doi.org/10.1016/j.palaeo.2010.05.037.

Pfretzschner, H.-U. (1998): Frühdiagenetische Prozesse bei der Fossilisation von Knochen, Neues Jahrbuch für Geologie und Paläontologie, Abhandlungen, 210(3): 369–397.

Pfretzschner, H.-U. (2004): Fossilization of Haversian bone in aquatic environments. Comptes Rendus Palevol, 3(6–7): 605–616. https://doi.org/10.1016/j.crpv.2004.07.006.

Prieto, J. (2007): Kleinsäuger-Biostratigraphie und Paläoökologie des höheren Mittelmiozäns (MN8) Bayerns: Spaltenfüllungen der Fränkischen Alb und Lokalitäten der Oberen Süßwassermolasse im Vergleich. PhD thesis, Fakultät für Geowissenschaften der Ludwig-Maximilians-Universität München, Munich, Germany.

Prieto, J. (2012): The Genus *Eomyops* Engesser, 1979 (Rodentia, Eomyidae) from the youngest deposits of the German part of the North Alpine Foreland Basin. Swiss Journal of Palaeontology, 131: 95–106. https://doi.org/10.1007/s13358-011-0033-4.

Prieto, J. & Dam, J.A. van (2012): Primitive Anourosoricini and Allosoricinae from the Miocene of Germany. Geobios, 45: 581–589. https://doi.org/10.1016/j.geobios.2012.03.001.

Prieto, J. & Rummel, M. (2009a): The genus *Collimys* Daxner-Höck, 1972 (Rodentia, Cricetidae) in the Middle Miocene fissure fillings of the Frankian Alb (Germany). Zitteliana, 48/49: 75–88.

Prieto, J. & Rummel, M. (2009b): Evolution of the genus *Collimys* Daxner-Hock, 1972 (Rodentia, Cricetidae) a key to Middle to Late Miocene biostratigraphy in Central Europe. Neues Jahrbuch für Geologie und Paläontologie – Abhandlungen, 252: 237–247. https://doi.org/10.1127/0077-7749/2009/0252-0237.

Prieto, J., Hoek Ostende, L.W. van den, Böhme, M. & Braze, M. (2011): Reappearance of *Galerix* (Erinaceomorpha, Mammalia) at the Middle to Late Miocene transition in South Germany: biostratigraphic and palaeoecologic implications. Contributions to Zoology, 80: 179–189. https://doi.org/10.1163/18759866-08003002.

Prieto, J., Angelone, C., Casanovas-Vilar, I., Gross, M., Hír, J., Hoek Ostende, L.W. van den, Maul, L.C. & Vasilyan, D. (2014): The small mammals from Gratkorn: an overview. Palaeobiodiversity and Palaeoenvironments, 94: 135–162. https://doi.org/10.1007/s12549-013-0147-3.

Schleich, H.H. (1981): Jungtertiäre Schildkröten Süddeutschlands unter besonderer Berücksichtigung der Fundstelle Sandelzhausen. Courier Forschungsinstitut Senckenberg, 48: 1–372.

Schleich, H.H. (1984): Neue Reptilienfunde aus dem Tertiär Deutschlands. 1. Schildkröten aus dem Jungtertiär Süddeutschlands. Naturwissenschaftliche Zeitschrift für Niederbayern, 30: 63–93.

Schneider, S. & Prieto, J. (2011): First record of an autochthonous community of fluviatile freshwater molluscs from the Middle / Late Miocene Upper Freshwater Molasse (southern Germany). Archiv für Molluskenkunde International Journal of Malacology, 140: 1–18. https://doi.org/10.1127/arch.moll/1869-0963/140/001-018.

Scholz, H. (2016): Bau und Werden der Allgäuer Landschaft – Alpen und schwäbisches Alpenvorland – zwischen Ammer und Bodensee. Eine süddeutsche Erd- und Landschaftsgeschichte. 3rd edition. Schweizerbart, Stuttgart: 354pp.

Schwerd, K., Doppler, H. & Unger, H.-J. (1996): Gesteinsfolge des Molassebeckens und der inneralpinen Tertiärbecken. – In: Freudenberger, W. & Schwerd, K. (Eds): Erläuterungen zur Geologischen Karte von Bayern 1:500.000. Bayerisches Geologisches Landesamt, München: 141–149.

Seitner, L. (1987): Miozäne Mikrofloren aus Sedimenten der Süssbrackwassermolasse und der Oberen Süsswassermolasse Süddeutschlands. PhD thesis, Fakultät für Geowissenschaften der Ludwig-Maximilians-Universität München, Munich, Germany.

Sinitsa, M.V., Čermák, S. & Kryuchkova, L.Y. (2021): Cranial Anatomy of *Csakvaromys bredai* (Rodentia, Sciuridae, Xerinae) and Implications for Ground Squirrel Evolution and Systematics. Journal of Mammalian Evolution, 29: 149–189. https://doi.org/10.1007/s10914-021-09561-w.

Steininger, F.F., Bernor, R.L. & Fahlbusch, V. (1989): European Neogene marine / continental chronologic correlations. – In: Lindsay, E.H., Fahlbusch, V. & Mein, P. (Eds): European Neogene mammal chronology, New York (Plenum): 15–46.

Uchman, A., Mikuláš, R., & Stachacz, M. (2016): Mayfly Burrows in Firmground of Recent Rivers from the Czech Republic and Poland, with Some Comments on Ephemeropteran Burrows in General. Ichnos, 24(3): 191–203. https://doi.org/10.1080/10420940.2016.1257488.

Ulmer-Scholle, D.S., Scholle, P.A., Schieber, J. & Raine, R.J. (2015): A color guide to the petrography of sandstones, siltstones, shales and associated rocks. American Association of Petroleum Geologists Memoir (Tulsa), 109: 526pp. https://doi.org/10.1306/M1091304.

Voorhies, M.R. (1969): Taphonomy and Population Dynamics of an Early Pliocene Vertebrate Fauna, Knox County, Nebraska. University of Wyoming, Laramie, WY. https://doi.org/10.2113/gsrocky.8.special_paper_1.1.

Wang, J., Yan, Y., Bai, J. & Su, X. (2020): Influences of riverbed siltation on redox zonation during bank filtration: a case study of Liao River, Northeast China. Hydrology Research, 51: 1478–1489. https://doi.org/10.2166/nh.2020.107.

Watson, J.A.L. & Abbey, H.M. (1986): The effects of termites (Isoptera) on bone: some archaeological implications. Sociobiology, 11: 245–254.

Weigelt, J. (1919): Geologie und Nordseefauna: Steinbruch, 14: 228–231, 244–246.

Weigelt, J. (1928): Rezente Wirbeltierleichen und ihre paläobiologische Bedeutung. 3rd edition, 1999, Dieter W. Berger, Bad Vilbel: 288pp.

Whittaker, R.J., Nogués-Bravo, D. & Araújo, M.B. (2007): Geographical gradients of species richness: a test of the water-energy conjecture of Hawkins et al. (2003) using European data for five taxa. Global Ecology and Biogeography, 16: 76–89. https://doi.org/10.1111/j.1466-8238.2006.00268.x.

Wiedenmann, L. (2011): Historischer Kohlenbergbau und Bergbauversuche auf Kohlen im Allgäu – Irsee. Eine Archiv-Dokumentation eines (fast) vergessenen Erwerbszweigs. Bauer-Verlag GmbH, Thalhofen: 373pp.

Williams, S.A., Prang, T.C., Meyer, M.R., Russo, G.A. & Shapiro, L.J. (2020): Reevaluating bipedalism in *Danuvius*. Nature, 586: E1–E3. https://doi.org/10.1038/s41586-020-2736-4.

Zhang, Y., Zhou, J., Lei, Z., Ren, H. & Chen, H. (2024): Effect of Acid Production from Forest Litter on the Availability of Heavy Metals in Soil. Forests, 15(12), 2097. https://doi.org/10.3390/f15122097.

Ziegler, R. (1999): Order Insectivora. – In: Rössner, G.E. & Heissig, K. (Eds): The Miocene Land Mammals of Europe. Verlag Dr. Friedrich Pfeil, München: 53–74.